AQA Physics for GCSE Combined Science: Trilogy

Higher Workbook

Helen Reynolds

Editor: Lawrie Ryan

OXFORD
UNIVERSITY PRESS

Great Clarendon Street, Oxford, OX2 6DP, United Kingdom

Oxford University Press is a department of the University of Oxford.
It furthers the University's objective of excellence in research,
scholarship, and education by publishing worldwide. Oxford is a
registered trade mark of Oxford University Press in the UK and in
certain other countries

British Library Cataloguing in Publication Data
Data available

978 0 19 837485 5

10 9 8 7 6 5 4 3 2

Paper used in the production of this book is a natural, recyclable
product made from wood grown in sustainable forests.
The manufacturing process conforms to the environmental regulations
of the country of origin.

Printed in Great Britain by CPI Group (UK) Ltd., Croydon CR0 4YY

Acknowledgements

Cover: Johnér/Offset

p9: Chones/Shutterstock; **p60**: Jaruek Chairak/Shutterstock;
p86: wang song/Shutterstock.

Artwork by Q2A Media Services Ltd.

Contents

Introduction

Key content – Each topic from your GCSE Student Book is covered, and includes a summary of the key content you need to know

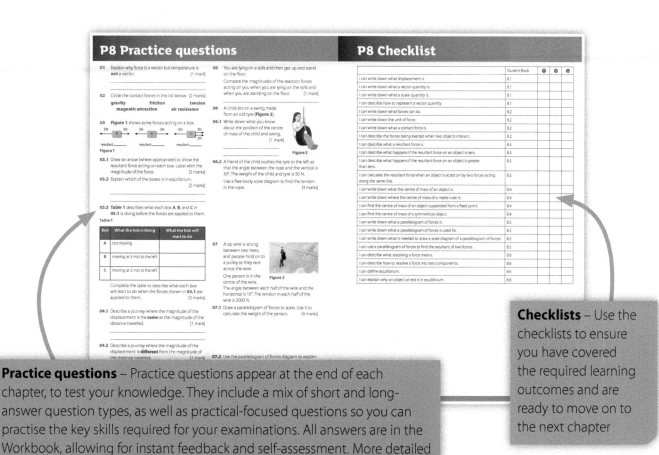

Practice activities – Lots of varied questions, increasing in difficulty, to build your confidence and help you progress through the course

Practice questions – Practice questions appear at the end of each chapter, to test your knowledge. They include a mix of short and long-answer question types, as well as practical-focused questions so you can practise the key skills required for your examinations. All answers are in the Workbook, allowing for instant feedback and self-assessment. More detailed exam-style feedback on practice questions is also available online

Checklists – Use the checklists to ensure you have covered the required learning outcomes and are ready to move on to the next chapter

P1.1 Changes in energy stores

A Fill in the gaps to complete the sentences.

An energy _____ is a way of keeping track of energy.

You can transfer energy by _____ , by _____ , by _____ , and by

_____ .

When an object falls, the energy in its _____ _____ energy store decreases, and the energy

in its _____ energy store increases.

When an object hits the ground but does not bounce, the energy in its _____ energy store

decreases. Energy is transferred to the _____ by sound waves and by _____ .

B There are different types of energy store. Give an example of an **activity** that you could do to make the change in each energy store described below.

a A decrease in the energy in the chemical store of a battery.

_____

_____

b An increase in the energy in a gravitational potential store.

_____

_____

c An increase in the energy in an elastic potential store.

_____

_____

C A wind-up torch produces light when you turn the handle.

Starting with when you eat food:

● describe the **changes in energy** in energy stores before and after you use the torch

● describe the **physical processes** that transfer energy between the stores.

_____

_____

_____

_____

_____

_____

_____

_____

P1.2 Conservation of energy

A Fill in the gaps to complete the sentences.

Energy cannot be _____ or _____ . This is the principle of _____ of energy, which applies to _____ energy changes.

An isolated system is called a _____ system. There are no _____ transfers into or out of the system.

If there are transfers within the system then the total energy does not _____ .

B On the Moon, there is no air. An astronaut on the Moon is holding a swinging pendulum. Energy transfers are taking place.

Suggest what will happen to the motion and height of the pendulum over time. Explain whether or not the pendulum is a closed system.

_____

_____

_____

_____

C What would happen if everyday situations became closed systems? Describe what you would observe in each of the situations below.

a A bouncing ball.

_____

_____

b A child on a swing in motion.

_____

_____

c A bungee jumper.

_____

_____

D There is a roller coaster at a funfair.

a Describe a system involving the rollercoaster that **is not** closed.

_____

_____

b Describe a system involving the rollercoaster that **is** closed.

_____

_____

P1.3 Energy and work

A Fill in the gaps to complete the sentences.

'Work' in science is about using a _____ to move an object. Work is a way of _____ energy between energy stores.

You can calculate work using this equation:

work done (_____) = _____

[You need to remember this equation.]

When an object moves through the air it does work against _____ , or when you slide it

across the floor it does work against _____ .

These processes _____ the surroundings.

B Calculate the work that a father does against friction when he pushes the buggy 180 m using a force of 15 N.

15 N

work done = _____

C Complete the table by calculating the work done by each force.

Force	Distance	Work
10 N	2 m	
30 N	10 cm	
25 kN	5 m	
20 kN	50 mm	

D A student uses a newton-meter to pull a tub full of sand across the floor. She wants to investigate how the mass of sand affects the work done against friction.

- Identify the independent, dependent, and control variables in this investigation.
- Describe and explain the measurements that the student needs to make and how she would use these measurements to find out how the mass of sand affects the work done against friction.

_____

_____

_____

_____

_____

_____

P1.4 Gravitational potential energy stores

A Fill in the gaps to complete the sentences.

The gravitational potential energy of an object _____ when it is lifted up and _____ when it is moved down.

You do _____ when you lift something up to overcome the gravitational _____ .

The gravitational field strength on the Moon is _____ than it is on the Earth, so it is _____ to lift an object up on the Moon than on the Earth.

You can calculate the change in the gravitational potential energy store using this equation:

gravitational potential energy (_____) = _____

[You need to remember this equation.]

B There is a change in the gravitational potential energy store when you lift a suitcase into a car. The suitcase has a mass of 30 kg. You lift it up 1 m. Gravitational field strength = 10 N/kg.

Calculate the change in the gravitational potential energy store. Remember to include units.

change in gravitational potential energy = _____

C Complete the table by calculating the change in gravitational potential energy for each object.

Mass of object	Gravitational field strength in N/kg	Change in height	Change in gravitational potential energy store in J
1 kg	10	2 m	
1 kg	1.6	10 cm	
250 g	27	5 m	
20 g	10	50 mm	

D In the investigation described in Topic **P1.3**, activity **D**, a student pulled a tub of sand across the floor. Now the student wants to investigate pulling the same tub of sand up a ramp.

- Compare the type of work done in the two investigations.
- The distance travelled in each investigation is the same. Explain which amount of work done would be bigger.

_____

_____

_____

_____

_____

_____

P1.5 Kinetic energy and elastic energy stores

A Fill in the gaps to complete the sentences.

The kinetic energy of an object depends on its _____ and its _____ .

You can calculate kinetic energy using this equation:

kinetic energy (_____) = _____

[You need to remember this equation.]

When you stretch or compress an object, you do _____ and transfer energy to an _____ store.

[You can use this equation on the Physics equation sheet to calculate the change in elastic potential energy:
elastic potential energy = 0.5 × spring constant (N/kg) × (extension (m))²]

B A student uses light gates to find the speed of a ball just before it hits the ground. He finds that the speed is 12.5 m/s. The mass of the ball is 50 g.

a Calculate the kinetic energy.

kinetic energy = _____

b The ball bounces. There is a point at which all the energy in the kinetic store of the ball is transferred to an elastic store. Write down what the ball is doing at this point.

_____

c Explain why the student needed to use light gates.

_____

_____

C A machine that throws tennis balls contains a spring with a spring constant of 145 N/m. When you push a ball into the machine, the spring compresses by 1.5 cm. The mass of the ball is 55 g.

Calculate the speed of the ball as it leaves the machine when you fire it.

speed of the ball = _____

P1.6 Energy dissipation

A Fill in the gaps to complete the sentences.

Useful energy is energy transferred in a pathway that we _____ .

In any device or process energy spreads out, which we call _____ .

Energy that is not useful is _____ . This energy is eventually transferred to the

_____ , which become _____ .

B For each of the activities below, describe **two** pathways that dissipate energy.

a Riding a bicycle.

_____

_____

b Using an electric drill.

_____

_____

c Using an electric kettle.

_____

_____

C On a car journey, there are lots of energy transfers. A student says that when you are travelling on a motorway at a steady speed of 60 mph, the energy in the chemical store of the fuel is being transferred to a kinetic store by the car's engine.

a Explain why this statement is wrong.

_____

_____

b Describe what happens to the energy in the fuel while a car is travelling at a steady speed.

_____

_____

c Suggest and explain whether the energy transfers that you have described in part **b** involve **useful energy** or **wasted energy**.

_____

_____

_____

_____

D You turn an electric heater on to heat a room. An hour later you turn it off. Describe this situation using the key words for this topic in the Student Book.

_____

_____

_____

_____

P1.7 Energy and efficiency

A Fill in the gaps to complete the sentences.

You can calculate efficiency using this equation:

efficiency = ────────────────────────

[*You need to remember this equation.*]

No device can be more than _____ % efficient, because this would mean that energy has been _____ .

Machines waste energy because of _____ between moving parts, by _____ the air when

they are moving, and by getting _____ when a current flows. You can reduce the amount of wasted

energy by _____ the surfaces between moving parts.

B a A motor has an input energy transfer of 1000 J and a useful output energy transfer of 400 J.
Calculate the **efficiency**.

efficiency = _____

b Calculate the input energy transfer required for the motor to transfer 25 J of useful output energy.

input energy transfer = _____

C Using an electric drill wastes a lot of energy. Suggest and explain how to improve the efficiency of a drill. Include **three** processes that waste energy and **two** ways to reduce the energy wasted.

_____

_____

_____

_____

_____

_____

D A student says that a ball that does not bounce very high is not very efficient. Is she correct? Explain why.

_____

_____

_____

_____

_____

P1.8 Electrical appliances

A Fill in the gaps to complete the sentences.

Most of the energy that people use in their homes is supplied by gas, _____ , or electricity.

_____ is a clean and efficient way of transferring energy to many of the appliances that you use every

day. You use electrical appliances for _____ (e.g., in an oven), _____ (e.g., a low-energy

lamp), moving objects (e.g., the turntable in a _____ oven), and creating sound and images.

More efficient electrical appliances waste _____ energy than less-efficient electrical devices.

B You can use an electric kettle or an oven to heat water. In each case, energy is transferred to the thermal store of the water.

Using the idea of efficiency, explain why you usually use a kettle and not an oven to heat water.

_____

_____

_____

C Here are some data about different types of light bulb. The bulbs appear equally bright.

Light bulb	Energy supplied to light bulb in 1 minute
X	700 J
Y	2500 J

a Identify the light bulb that is more efficient. _____

b Explain why you did not need to do a calculation to work out the answer to part **a**.

_____

_____

c Identify the light bulb that wastes more energy. _____

d Explain your answer to part **c**.

_____

_____

D You are shopping for an electrical appliance. To help you choose between different appliances, you compare the different prices.

Explain why you also need to compare power and efficiency when choosing an appliance.

_____

_____

_____

_____

_____

_____

_____

_____

P1.9 Energy and power

A Fill in the gaps to complete the sentences.

Power is the _____ of energy transfer.

You can calculate power using this equation:

$$\text{power} = \frac{\underline{\hspace{3cm}} (\underline{\hspace{1cm}})}{\underline{\hspace{3cm}} (\underline{\hspace{1cm}})}$$ **[You need to remember this equation.]**

You can calculate efficiency as a percentage using this equation:

$$\text{percentage efficiency} = \frac{\overline{\underline{\hspace{6cm}}}}{\underline{\hspace{6cm}}} \quad \underline{\hspace{2cm}}$$

[You need to remember this equation.]

You can calculate the power that is wasted using this equation:

power wasted = _____ power in – _____ power out

B A student says that more powerful devices are more efficient.

Explain why this is not always true.

_____

_____

C A lift carries you to the top floor of a building, transferring 170 kJ of energy in 15 seconds.

a Calculate the useful power output of the lift.

   useful power = _____ W

b The total power in is 20 kW.

   Calculate the percentage efficiency of the lift.

   efficiency = _____ %

c Calculate the wasted power.

   wasted power = _____ W

P1 Practice questions

01 You drop a ball and it bounces as shown in **Figure 1**.

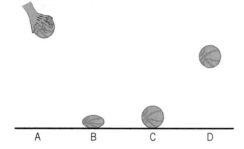

A	B	C	D

Figure 1

01.1 Describe the changes in energy between points **A** and **B**. [2 marks]

_____

_____

01.2 Explain why the ball reaches a lower height when it bounces. [1 mark]

_____

_____

02 A scientist measures the time it takes two kettles with the same power rating to boil water. **Table 1** shows the data from the experiment.

Table 1

Kettle	Energy required to boil water in J	Energy supplied to kettle in J
A	300 000	400 000
B	300 000	450 000

02.1 Describe how the scientist calculated the energy suppled to the kettle. [2 marks]

_____

_____

02.2 Identify **one** control variable in this experiment. [1 mark]

_____

02.3 Calculate the efficiency of kettle **A**. Write down the equation that you will use. [3 marks]

_____

_____

_____

02.4 A manufacturer's kettles all have the same power rating. The manufacturer wants to make one kettle more efficient than the rest. Suggest how it could do this. [1 mark]

_____

_____

03 About 100 years ago there was an event in the Olympic Games called the standing high jump. An athlete stood still, then jumped vertically. The world record is 1.9 m.

03.1 If the mass of the athlete is 50 kg, and gravitational field strength is 10 N/kg, calculate the take-off speed of the athlete. Write down the equations that you will use. [4 marks]

_____

_____

_____

_____

_____

03.2 Just before landing, the athlete has a speed of 6 m/s. Explain the difference between the take-off and landing speeds. [1 mark]

_____

_____

04 **Table 2** shows some data about the lifts in two very tall buildings. Compare the lifts when each lift is carrying 10 people, each with a mass of 70 kg, from the ground floor to the top of each building. Use ideas about work, power, and useful and wasted energy.

Gravitational field strength = 10 N/kg. [6 marks]

Table 2

	Empire State Building	The Shard
number of floors to observation floor	86	70
height of one floor	3 m	3 m
time to reach the observation floor	55 seconds	1 minute
efficiency of lift motor	85%	90%

_____

_____

_____

_____

_____

_____

_____

_____

_____

_____

_____

_____

P1 Checklist

	Student Book	☺	☺	☹
I can describe the ways in which energy can be stored.	1.1			
I can describe how energy can be transferred.	1.1			
I can describe the energy transfers that happen when an object falls.	1.1			
I can describe the energy transfers that happen when a falling object hits the ground without bouncing back.	1.1			
I can describe what conservation of energy is.	1.2			
I can explain why conservation of energy is a very important idea.	1.2			
I can describe what a closed system is.	1.2			
I can describe energy transfers in a closed system.	1.2			
I can describe what work means in science.	1.3			
I can describe how work and energy are related.	1.3			
I can calculate the work done by a force.	1.3			
I can describe what happens to work that is done to overcome friction.	1.3			
I can describe what happens to the gravitational potential energy store of an object when it moves up or down.	1.4			
I can explain why an object moving up increases its gravitational potential energy store.	1.4			
I can explain why it is easier to lift an object on the Moon than on the Earth.	1.4			
I can calculate the change in gravitational potential energy of an object when it moves up or down.	1.4			
I can write down what the kinetic energy of an object depends on.	1.5			
I can calculate kinetic energy.	1.5			
I can describe what an elastic potential energy store is.	1.5			
I can calculate the amount of energy in an elastic potential energy store.	1.5			
I can describe what is meant by useful energy.	1.6			
I can describe what is meant by wasted energy.	1.6			
I can describe what eventually happens to wasted energy.	1.6			
I can describe if energy is still as useful after it is used.	1.6			
I can describe what is meant by efficiency.	1.7			
I can write down the maximum efficiency of any energy transfer.	1.7			
I can describe how machines waste energy.	1.7			
I can describe how energy is supplied to homes.	1.8			
I can explain why electrical appliances are useful.	1.8			
I can describe what most everyday electrical appliances are used for.	1.8			
I can explain how to choose an electrical appliance for a particular job.	1.8			
I can describe what is meant by power.	1.9			
I can calculate the power of an appliance.	1.9			
I can calculate the efficiency of an appliance in terms of power.	1.9			
I can calculate the power wasted by an appliance.	1.9			

P2.1 Energy transfer by conduction

A Fill in the gaps to complete the sentences.

The best conductors of energy are _____ . Materials such as wool and fibreglass, which are

_____ , are good insulators.

A material with a high thermal conductivity has a _____ rate of energy transfer through it.

The thicker a layer of insulating material, the _____ the rate of energy transfer through it.

B The diagram shows an experiment to investigate conduction. The wax on rod **X** melts first, then the wax on rod **Z**, and finally the wax on rod **Y**.

a List the rods in order of thermal conductivity, starting with the **lowest**.

_____

b Describe how you worked out the answer to part **a**.

_____

_____

c Suppose you had two rods with similar thermal conductivities. Suggest and explain **two** improvements to this experiment that would help you tell the difference between the two rods.

_____

_____

_____

_____

C Loft insulation needs to trap lots of air. Explain what this tells you about the thermal conductivity of the loft insulation material compared with the thermal conductivity of air.

_____

_____

_____

_____

P2.2 Specific heat capacity

A Fill in the gaps to complete the sentences.

The specific heat capacity is the energy needed to change the temperature of _____ of a substance by

_____ °C.

Using a heater with the same rate of energy transfer, a more massive piece of a substance will take _____ to heat up than a less massive piece of the same substance.

To find the specific heat capacity you need to measure the _____ using a joulemeter, the temperature

difference using a _____ , and the _____ using a digital balance.

[*You can use this equation on the Physics equation sheet to calculate the change in thermal energy:*
change in thermal energy (J) = mass (kg) × specific heat capacity (J/kg °C) × change in temperature (°C)]

B Look at the amounts of energy in the table below.

The specific heat capacity of aluminium is 900 J/kg °C, and the specific heat capacity of water is 4200 J/kg °C.

	Energy required to...
X	raise the temperature of 2 kg of aluminium by 20 °C
Y	raise the temperature of 1 kg of water by 20 °C
Z	raise the temperature of 1 kg of water by 10 °C

Write the letters in order from the smallest to the largest amount of energy, and explain how you worked out the order:

_____

_____

_____

C The samples in activity **B** are heated with an immersion heater with a power of 100 W. Compare the times taken to heat samples **X** and **Z**.

_____

_____

D A student puts 0.25 kg of water in an insulated beaker. She puts a heater in the beaker and switches it on. It raises the temperature of the water by 11.5 °C in 15 minutes.

a Calculate the change in thermal energy.

The specific heat capacity of water is 4200 J/kg °C.

change in thermal energy = _____ J

b Suggest what the student would notice if she used twice as much water.

_____

P2.3 Heating and insulating buildings

A Fill in the gaps to complete the sentences.

People use heaters that run on electricity or _____ to heat their homes, or central heating that runs on

_____ or gas. Solid fuel such as coal or _____ is often burned in stoves for heating.

People can reduce the energy transfer from the loft of a house using _____ _____ .

They can reduce the rate of energy transfer through windows using _____ _____ .

There are usually _____ layers of brick in the walls of a house, with a layer of _____

_____ insulation to reduce the rate of energy transfer. It also helps to use bricks on the outside that are

_____ and have a _____ thermal conductivity.

B Here are some fuels and some energy transfer devices used for heating in people's homes.

Tick the boxes to show which fuels each type of device may use. Each device may use more than one fuel.

Fuel	✓ if it may be burned in a stove	✓ if it may be used in a central heating system	✓ if it may be burned on a fire
oil			
coal or wood			
gas			

C a Describe what cavity wall insulation is, and refer to its thermal conductivity.

_____

_____

b Name **one** other method of insulating a house that works the **same** way as cavity wall insulation.
Explain your choice.

_____

_____

c Name **one** method of insulating a house that works in a **different** way to cavity wall insulation.
Explain your choice.

_____

_____

D A family is deciding between installing double glazing or loft insulation.

Double glazing costs £2000, and would save the family £100 per year on their heating bills.

Loft insulation costs £175, and would save them £25 per year on their heating bills.

They want the one with the lowest costs over a 30-year period. Which would you recommend that they choose?

_____

_____

_____

_____

_____

P2 Practice questions

01 A student puts an equal mass of water at 70 °C into three cans. She puts insulation around each can. She uses the same type of insulation, but different thicknesses. She measures the temperature of the water 15 minutes later. Her results are shown in **Table 1**.

Table 1

	Temperature after 15 minutes in °C
X	40
Y	55
Z	52

01.1 Write the letter of the can with the thickest insulation. _____ [1 mark]

01.2 Write the letter of the can from which the least energy was transferred to the surroundings.
_____ [1 mark]

01.3 Suggest and explain **one** other variable that the student needs to control in this experiment.
[2 marks]

_____

_____

02 **Figure 1** shows some equipment that you could use to measure the specific heat capacity of a substance.

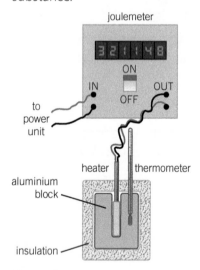

Figure 1

02.1 Explain why you need to insulate the aluminium block. [1 mark]

_____

_____

02.2 A student makes the following measurements using an insulated aluminium block.

temperature rise = 20 °C

mass of block = 1 kg

energy transferred = 45 J

Use these measurements to calculate the specific heat capacity of aluminium. Use this equation:

specific heat capacity (J/kg °C) =

$$\frac{energy\ (J)}{mass\ (kg) \times temperature\ change\ (°C)}$$ [2 marks]

_____

_____

_____

_____

02.3 The student does the same experiment, but this time does not insulate the block. Explain what would happen to the student's measurement of the specific heat capacity of the uninsulated block.
[2 marks]

_____

_____

03 In an article published in 1955, a scientist suggested that keeping someone at a comfortable sleeping temperature inside a sleeping bag depends on the thickness of the insulation of the bag.

Table 2

Temperature outside the sleeping bag in °C	Temperature difference between human body and outside the sleeping bag in °C	Thickness of the sleeping bag insulation in mm
4	33	38
−7	44	51
−18	55	64
−29	66	76
−40	77	89

03.1 Use the data from **Table 2** to show that the thickness of insulation needed for the sleeping bag is **proportional** to the difference between the temperature of the body inside the sleeping bag and the temperature outside the bag. [2 marks]

_____

_____

03.2 Explain why a modern sleeping bag designed for use at −40 °C is much thinner than 89 mm.
[2 marks]

_____

_____

_____

_____

P2 Checklist

	Student Book	😊	😐	☹
I can write down which materials make the best conductors.	2.1			
I can write down which materials make the best insulators.	2.1			
I can describe how the thermal conductivity of a material affects the rate of energy transfer through it by conduction.	2.1			
I can describe how the thickness of a layer of material affects the rate of energy transfer through it by conduction.	2.1			
I can describe what the specific heat capacity of a substance means.	2.1			
I can calculate the energy needed to change the temperature of an object.	2.2			
I can describe how the mass of a substance affects how quickly its temperature changes when you heat it.	2.2			
I can describe how to measure the specific heat capacity of a substance.	2.2			
I can describe how homes are heated.	2.2			
I can describe how you can reduce the rate of energy transfer from your home.	2.3			
I can describe what cavity wall insulation is.	2.3			

P3.1 Energy demands

A Fill in the gaps to complete the sentences.

We meet most of our energy demands by burning _____, _____, and _____ .

These energy resources are non-renewable which means that they _____ run out.

Renewable resources _____ _____ run out.

Renewable fuels from living or recently living material are called _____ . One example is a gas called

_____ , and another example is a liquid called _____ .

We use uranium or plutonium in a _____ power station. These fuels release much _____ energy per kilogram than fossil fuels.

B We use energy resources for heating, transportation, and generating electricity.

a Suggest **one** fuel that could be used for all three purposes. _____

b Suggest **one** fuel that is used mainly for heating. _____

C a Describe what is meant by a biofuel.

_____

_____

b Explain why a biofuel is called renewable.

_____

_____

D a Explain why uranium atoms can be used to generate energy in nuclear power stations.

_____

b We can generate electricity with nuclear fuel or biofuels. The table below shows some information about the energy released by each kilogram of uranium that undergoes fission (fissile uranium) and a biofuel.

Fuel	Energy released per kg in MJ
fissile uranium	77 000 000
biofuel (animal manure)	12

i Calculate the mass of fissile uranium needed to release the same amount of energy as 1 tonne (1000 kg) of animal manure.

_____

_____

_____

ii Only 0.7% of uranium metal is fissile.

Calculate the mass of uranium metal that is needed to release the same amount of energy as 1 kg of animal manure.

_____

_____

_____

_____

P3.2 Energy from wind and water

A Fill in the gaps to complete the sentences.

A wind _____ is an electricity generator on top of a tall tower.

A _____ generator is used to generate electricity from wave power.

Water stored in lakes or reservoirs can run downhill, flowing through _____ that turn generators. This is

called _____ power.

In a _____ power station, water at high tide is trapped behind a barrage, then released to turn a generator.

Wind and wave power can be unreliable, and renewable resources can damage the _____ .

B a Both geothermal power and tidal power are renewable. Describe **one** similarity between geothermal power
and tidal power, using the idea of how electricity is generated.

_____

_____

b Both wind power and wave power are renewable. Describe **one** similarity between wind power and wave
power, using the idea of how electricity is generated.

_____

_____

C One of the first hydroelectric power stations was at Niagara Falls in the USA.

Describe what happens in a hydroelectric power station.

_____

_____

_____

D The table shows the total cost per MW for different power stations that
use water.

Use the information in the table, and what you have learnt about the cost
of building power stations, to suggest **one** reason for each of the following
statements.

Power station	Total cost in £ per MW
hydroelectric	50
tidal	300
wave	100

a Hydroelectric power is cheapest: _____

_____

_____

_____

b Tidal power is more expensive than wave power:

_____

_____

_____

P3.3 Power from the Sun and the Earth

A Fill in the gaps to complete the sentences.

We use solar cells to _____ _____. They produce _____ amounts of

electricity, so you need lots of them. They are _____ to buy, and cost _____ to run.

We use solar heating panels to _____ water directly.

A solar power tower uses _____ to focus sunlight onto a water tank to produce steam, which can

generate _____.

Deep in the Earth, energy is released by _____ substances. This heats _____ that is pumped

deep down into the rocks. The water turns to _____, which drives _____ at the Earth's surface

to generate electricity.

B Compare a solar heating panel, a solar cell panel, and a solar power tower.

_____

_____

_____

_____

C Describe **two** advantages and **four** disadvantages of using solar cells to generate electricity.

_____

_____

_____

_____

_____

_____

D a A student makes a statement about power stations. State and explain whether you agree.

'A nuclear power station is like a geothermal power station because both use radioactivity.'

_____

_____

_____

b Explain why geothermal power stations may not be an alternative to fossil fuel power stations in
some locations.

_____

_____

_____

P3.4 Energy and the environment

A Fill in the gaps to complete the sentences.

Burning fossil fuels releases _____ gases, which could cause global _____ . It also releases

sulfur dioxide, which can produce _____ _____ .

Nuclear fuels produce _____ energy per kilogram of fuel than fossil fuels, but also produce

_____ waste. Nuclear power stations are _____ to decommission, and dangerous if there is

an accident.

Renewable energy resources _____ _____ produce harmful waste products, and they can be

used in _____ places. However, they can take up a large area and disturb the habitats of _____ and

_____ . They can be _____ to manufacture or install.

B Describe **two** ways to reduce the environmental impact of coal-fired power stations.

_____

_____

C The table lists some disadvantages of renewable energy sources.

Tick the correct columns to show which renewable resources each disadvantage applies to. You may need to tick several columns for each disadvantage.

Disadvantage	Wind?	Tidal?	Hydro?	Solar?
can cause noise pollution				
can affect river estuaries and the habitats of plants and animals there				
depends on the weather to work				
involves large reservoirs of water, which can affect the habitats of plants and animals				
needs large areas of land to produce enough energy from these panels				
not always available on demand				

D An island community wants to build a single power station to meet its electricity demand. Explain each of the following, using the idea of the environment:

a why a nuclear power station would be better than a fossil fuel power station.

_____

_____

_____

_____

b why a fossil fuel power station would be better than a nuclear power station.

_____

_____

_____

_____

A Fill in the gaps to complete the sentences.

The demand for electricity varies over the day, and during the year. You can meet this demand with _____

fired power stations and _____ storage. Nuclear power stations are _____ to build, run, and

_____ (dismantle when you no longer need them).

The carbon dioxide produced by burning fuels can be removed from the atmosphere in a process called

_____ _____, but this is very expensive. Renewable resources are _____ to run

but _____ to install. We are going to need a range of resources to meet future demand for energy.

B Look at the graph showing the demand for electricity during a typical day.

a Suggest what 'base load' means.

_____

_____

_____

b Suggest how excess energy is stored.

_____

_____

_____

c Explain why this graph indicates that solar power cannot be relied upon to meet demand during the day.

_____

_____

C a Describe the difference between the capital costs and the running costs of a power station.

_____

_____

_____

b Suggest what might happen in the future to the capital costs of renewable energy resources.

_____

_____

D The capital costs of a nuclear power station and an offshore wind farm are about the same, but the total costs of the nuclear power station are half the total costs of the wind farm. Suggest why.

_____

_____

_____

_____

P3 Practice questions

01 Our energy demands can be met using a variety of fuels and energy resources.

01.1 Write down the **three** main fuels that we use today. [1 mark]

_____

01.2 Electricity can be generated using wind, waves, and tides. Tick the columns in **Table 1** that apply to each of these resources. You may need to tick more than one column for each resource. [3 marks]

Table 1

Type of resource	Uses water to turn a turbine	Uses air to turn a turbine	A turbine turns a generator	Reliable resource
wind				
waves				
tides				

02 Starting with uranium fuel rods, describe how a nuclear power station generates electricity. [6 marks]

_____
_____
_____
_____
_____
_____

03 **Figure 1** is a pie chart that shows the sources we currently use to generate electricity in the UK.

Figure 1

other fuels 5% and renewables
oil 1%
hydro 1%
nuclear 16%
gas 46%
coal 31%

03.1 Calculate the percentage of the UK's electricity that comes from fossil fuels. [2 marks]

_____
_____

03.2 Compare and contrast the effects on the environment of burning biofuels and fossil fuels. [4 marks]

_____
_____
_____
_____

03.3 Name an energy source in **Figure 1** that is in the category 'renewables'. [1 mark]

_____

03.4 Write down **one** advantage and **one** disadvantage of using the energy source named in **03.3**. [2 marks]

_____
_____
_____

04 Dev and Sasha are arguing about which energy resources we should use in the future to generate electricity.
Write down the answer each person would give to the other. Include **two** reasons in each box. [4 marks]

Dev says: I think we should use nuclear power. It is reliable and doesn't produce greenhouse gases.	Sasha's reply:
Sasha says: I think we should use renewables. They are cheaper and better for the environment.	Dev's reply:

05 Solar panels save energy.

05.1 Give the **two** types of solar panel you can use on the roof of a house. [2 marks]

_____

05.2 There are about 25 million houses in the UK, and the area of each roof is about 140 m². The maximum output of a solar cell on a roof is about 250 W/m². Calculate the percentage of UK houses that would need to cover their roofs in solar cells to meet the UK base load of 5500 MW. [4 marks]

_____
_____
_____
_____
_____
_____

05.3 Comment on the assumptions that you used in your calculation in **05.2**. Include an evaluation of the feasibility of powering the UK using solar cells. [4 marks]

_____
_____
_____
_____

P3 Checklist

	Student Book	☺	☺	☹
I can describe how most energy demands are met today.	3.1			
I can name the energy resources that are used.	3.1			
I can describe how nuclear fuels are used in power stations.	3.1			
I can name the other fuels that are used to generate electricity.	3.1			
I can describe what a wind turbine is made up of.	3.2			
I can describe how waves can be used to generate electricity.	3.2			
I can name the type of power station that uses water running downhill to generate electricity.	3.2			
I can describe how the tides can be used to generate electricity.	3.2			
I can describe what solar cells are and how they are used.	3.3			
I can describe the difference between a panel of solar cells and a solar heating panel.	3.3			
I can describe what geothermal energy is.	3.3			
I can describe how geothermal energy can be used to generate electricity.	3.3			
I can describe what fossil fuels do to the environment.	3.4			
I can explain why people are concerned about nuclear power.	3.4			
I can describe the advantages and disadvantages of renewable energy resources.	3.4			
I can evaluate the use of different energy resources.	3.4			
I can describe how best to use electricity supplies to meet variations in demand.	3.5			
I can compare the economic costs of different energy resources.	3.5			
I can name energy resources that need to be developed to meet people's energy needs in the future.	3.5			

P4.1 Current and charge

A Fill in the gaps to complete the sentences.

Every circuit component has its own circuit _____ , and you use these to draw circuit diagrams.

A battery consists of two or more _____ .

Current is the _____ of flow of charge. You can calculate current using this equation, with units in the brackets:

$$\text{current (\_\_\_)} = \frac{\overline{\rule{6cm}{0pt}}}{\rule{6cm}{0pt}}$$

[You need to remember this equation.]

B Next to each the name of each circuit component below, draw the circuit symbol.

diode	
fuse	
cell	

resistor	
variable resistor	
battery	

C All of the following statements are false. Below each one write the correct statement.

a Charge is measured in amperes.

_____

_____

b Time is always measured in minutes.

_____

_____

c Current gets smaller further away from the battery.

_____

_____

d Current is the flow of charge.

_____

_____

D In a torch lamp, a charge of 15 C flows in 2 minutes. Calculate the current.

current = _____ A

P4.2 Potential difference and resistance

A Fill in the gaps to complete the sentences.

Potential difference is the _____ transferred to each charge by the battery, or the _____

transferred by each charge to the circuit component. Potential difference is measured in _____.

You can calculate potential difference and resistance using these equations:

$$\text{potential difference (\_\_\_)} = \frac{\rule{5cm}{0.4pt}}{\rule{5cm}{0.4pt}}$$

$$\text{resistance (\_\_\_)} = \frac{\rule{4cm}{0.4pt}}{\rule{4cm}{0.4pt}}$$

[You need to remember these equations.]

Ohm's law says that the current through a resistor is _____ _____ to the potential difference

across it. If you reverse the potential difference across a resistor, you _____ the current through it.

B a Sketch a circuit with a battery, lamp, ammeter, and voltmeter showing how to measure the current through the lamp and the potential difference across it.

b Explain the position of the ammeter and the voltmeter in the circuit.

_____

_____

c Suggest and explain what would happen to the readings on the ammeter if you reversed the battery.

_____

_____

C A charge flows through a resistor and transfers 200 J of energy. If the potential difference across it is 12 V, calculate the charge that flows through the resistor.

charge = _____ C

D a The resistor in activity **C** has a resistance of 30 Ω. Calculate the time it would take for 600 C to flow through it.

time = _____ s

b State **one** assumption that you made about the resistor in part **a**.

_____

P4.3 Component characteristics

A Fill in the gaps to complete the sentences.

The resistance of a filament bulb _____ if the temperature increases.

For a diode the resistance in the forward direction is _____ and the resistance in the reverse direction is

_____ . A light-emitting diode (LED) emits light when a current passes through it in the _____

direction.

If the temperature of a thermistor increases, its resistance _____ . If the light intensity on a light-

dependent resistor (LDR) increases, its resistance _____ .

B Here are three graphs of current against potential difference (p.d.).

Component **X**	Component **Y**	Component **Z**

a Write the letter of the component that is an ohmic conductor, and explain your choice.

_____

_____

b Write the letter of the component that has a resistance that increases with p.d., and explain your choice.

_____

_____

c Name the remaining component, and describe how its resistance changes with p.d.

_____

_____

C Complete the table about light-dependent resistors and thermistors.

Device	Has a large resistance when it is …	Has a small resistance when it is …
light-dependent resistor		
thermistor		

D Explain why the resistance of a filament lamp increases with potential difference.

_____

_____

P4.4 Series circuits

A Fill in the gaps to complete the sentences.

In a series circuit:

- the _____ is the same in each component
- the total _____ _____ is shared between the components
- you find the total resistance by _____ the resistance of all the components.

If you have more than one cell in series, then you _____ all the potential differences to find the total potential difference.

If you add more resistors in series, the total resistance _____ . This is because the current through the resistors is _____ but the total potential difference across them is the same.

B Here is a series circuit.

a Complete the tables about the current and potential difference in the circuit. The bulbs are identical.

Position	Current in A
X	0.2
Y	
Z	

Component	Potential difference across it in V
cell	3
bulb 1	
bulb 2	

b Explain how you completed the tables.

_____

_____

_____

_____

C Calculate the resistance of each bulb in activity **B**, and the total resistance of the circuit.

resistance of each bulb = _____

total resistance of the circuit = _____

D A student removes a bulb from the circuit in activity **A**. Explain why this may not double the current in the circuit.

_____

_____

P4.5 Parallel circuits

A Fill in the gaps to complete the sentences.

In a parallel circuit:

- the _____ _____ across each component is the same

- you find the total _____ by adding the _____ through each component.

If you use a component that has a bigger resistance, the current through it will be _____ .

You can calculate current using this equation:

$$\text{current } (\underline{\quad}) = \frac{\overline{}}{}$$

[You need to remember this equation.]

If you add more resistors in parallel, the total resistance _____ because the total current _____ but the potential difference is the same.

B Here is a **parallel** circuit.

Position	Current in A
X	0.2
Y	
Z	

Component	Potential difference across it in V
cell	9
bulb 1	
bulb 2	

a Complete the tables about the current and potential difference in the circuit. The bulbs are identical.

b Explain how you completed the tables.

_____

_____

_____

_____

C Calculate the resistance of each bulb in activity **B**, and the total resistance of the circuit.

resistance of each bulb = _____

total resistance of the circuit = _____

D A student wants to increase the current through bulb 1 in the circuit in activity **B**. Describe **one** change she can make to the circuit to achieve this.

_____

_____

P4 Practice questions

01 A student connects the circuit in **Figure 1** to collect data to plot a characteristic curve for a lamp.

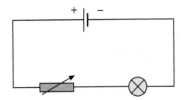

Figure 1

01.1 Name the two other circuit components that the student needs to be able to collect the data.
[2 marks]

_____

01.2 Explain why the student needs to include a variable resistor in the circuit.   [1 mark]

_____
_____

02 A student is looking at two mystery circuits on the bench labelled circuit **A** and circuit **B**.

- Each circuit contains a battery pack, two bulbs, and two switches.
- All of the wires connecting the bulbs to the switches and battery have been covered up by black paper.
- The student knows that each switch is connected next to a bulb.
- All the bulbs are on.
- He knows that one of the circuits is wired in series and the other is wired in parallel.

02.1 Describe how the student can work out which circuit is in series and which is in parallel.   [1 mark]

_____
_____

02.2 The resistance of one bulb is $10\,\Omega$, and the resistance of the other bulb is $15\,\Omega$. Calculate the resistance of the series circuit.   [1 mark]

_____
_____

02.3 The p.d. of the battery in each circuit is 12 V. Calculate the current through each bulb in the parallel circuit.   [2 marks]

_____
_____

02.4 Calculate the total resistance of the parallel circuit.
[3 marks]

_____
_____

03 A battery-operated fan spins when you turn it on.

03.1 A charge of 20 C flows through the fan in 40 seconds. Show that the current is 0.5 A.   [2 marks]

_____
_____

03.2 Write down Ohm's law.   [2 marks]

_____
_____

03.3 The potential difference across the fan is 9 V. Calculate the resistance.   [2 marks]

_____
_____

03.4 Write down what would happen to the fan if the potential difference across the fan was reversed.
[1 mark]

_____
_____

03.5 Use the definitions of current and p.d. to explain why doubling the p.d. of the power supply to the fan multiplies the energy transferred to the fan per second by a factor of 4.   [3 marks]

_____
_____

04 A student wants to monitor the light levels that a plant is receiving. She sets up the circuit in **Figure 2**.

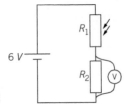

Figure 2

The resistance of the light-dependent resistor (LDR), R_1, is $100\,\Omega$ in the light and $500\,000\,\Omega$ in the dark. The resistance R_2 is $500\,\Omega$.

04.1 Explain why the light levels affect the reading on the voltmeter.   [3 marks]

_____
_____

04.2 Calculate and compare the reading on the voltmeter when the LDR is in the light, and the dark.   [6 marks]

_____
_____
_____
_____
_____
_____
_____

P4 Checklist

	Student Book	☺	☻	☹
I can describe how electric circuits are shown as diagrams.	4.1			
I can write down the difference between a battery and a cell.	4.1			
I can describe what determines the size of an electric current.	4.1			
I can calculate the size of an electric current from the charge flow and the time taken.	4.1			
I can write down what is meant by potential difference.	4.2			
I can write down what resistance is and what its unit is.	4.2			
I can write down Ohm's law.	4.2			
I can describe what happens when you reverse the potential difference across a resistor.	4.2			
I can describe what happens to the resistance of a filament lamp as its temperature increases.	4.3			
I can describe how the current through a diode depends on the potential difference across it.	4.3			
I can describe what happens to the resistance of a temperature-dependent resistor as its temperature increases.	4.3			
I can describe what happens to the resistance of a light-dependent resistor as the light level increases.	4.3			
I can describe the current, potential difference, and resistance for each component in a series circuit.	4.4			
I can describe the potential difference of several cells in series.	4.4			
I can calculate the total resistance of two resistors in series.	4.4			
I can explain why adding resistors in series increases the total resistance.	4.4			
I can describe the currents and potential differences for components in a parallel circuit.	4.5			
I can calculate the current through a resistor in a parallel circuit.	4.5			
I can explain why the total resistance of two resistors in parallel is less than the resistance of the smaller individual resistor.	4.5			
I can explain why adding resistors in parallel decreases the total resistance.	4.5			

P5.1 Alternating current

A Fill in the gaps to complete the sentences.

Direct current (d.c.) flows in _____ direction. Alternating current (a.c.) _____ its direction of flow.

In a mains circuit there is a live wire. Its potential difference alternates between _____ and

_____ every cycle. There is also a neutral wire. Its potential difference is _____ volts.

The National Grid is a _____ of cables and transformers that supply electricity to your home.

The peak potential difference of an a.c. supply is the _____ potential difference measured from

_____ volts. You can find the frequency of an a.c. supply by measuring the _____

_____, and using the equation:

$$\text{frequency} \, (\underline{}) = \frac{\rule{5cm}{0.4pt}}{\rule{5cm}{0.4pt}}$$

B Compare a graph of potential difference against time for a d.c. circuit and for an a.c. circuit.

_____

_____

C Look at the graph.

a Write down the peak potential difference.

_____

b Explain how to use the graph to find the frequency of the a.c.

_____

_____

c Calculate the frequency of the p.d. shown on the graph.

_____

_____

d State and explain whether the graph shows the p.d. of the live wire or the p.d. of the neutral wire.

_____

_____

e Suggest why we say that mains voltage is 230 V, and not the peak voltage that is shown on the graph.

_____

P5.2 Cables and plugs

A Fill in the gaps to complete the sentences.

Sockets and plugs are made of stiff _____ that encloses electrical connections. This material is used

because it is a good _____ .

A mains cable is made up of two or three insulated wires made of _____ surrounded by an outer layer of

flexible _____ material.

In a three-pin plug or a three-core cable, the insulation on the live wire is coloured _____, the neutral wire

is coloured _____, and the earth wire is coloured _____ and _____ .

The earth wire is connected to the _____ pin in a plug. It is used to earth the metal _____

of a mains appliance.

B **a** Complete the table showing the colours, functions, and potential differences of each of the wires in a
mains cable.

Name of wire in a mains cable	Colour	Function	Potential difference in V
live		carries the current to make an appliance work	
	blue		
earth			0

b Describe **one** similarity and **one** difference between the material insulating the wires inside a plug and the
material of the plug casing.

_____

_____

C **a** Describe how you can get an electric shock from an appliance.

_____

_____

b Explain how the fuse and earth wire protect you from an electric shock.

_____

_____

_____

c Suggest an appliance that does not need an earth wire. Explain your choice.

_____

_____

d Suggest and explain what would happen if you connected an appliance using just the earth and
neutral wires.

_____

_____

P5.3 Electrical power and potential difference

A Fill in the gaps to complete the sentences.

Power is the _____ transferred per second.

You can calculate energy transferred, the power, and the fuse rating using these equations:

energy transferred (_____) = _____

electrical power (_____) = _____

fuse rating (_____) = $\dfrac{\rule{3.5in}{0pt}}{\rule{3.5in}{0pt}}$

[You need to remember these equations.]

B a Define power.

_____

_____

b All mains appliances use the same potential difference.

Describe and explain the link between the power rating of an appliance and the fuse rating it needs.

_____

_____

_____

C An electric car has an electric motor with a power of 15 kW.

Calculate the energy transferred during a 2 ½ hour car journey.

energy transferred = _____ J

D a Calculate the current in a mains (230 V) microwave that has a power of 1000 W.

current = _____ A

b The fuses available are 1 A, 3 A, 5 A, and 13 A. Write down the fuse that you need for the microwave oven in part **a**. Explain your choice.

_____

_____

E Calculate the resistance of a mains (230V) games console. It needs a current of 1.5 A. It has a power of 350 W.

resistance = _____ Ω

P5.4 Electrical currents and energy transfer

A Fill in the gaps to complete the sentences.

You can calculate charge flow and the energy transferred using these equations, with units in the brackets:

charge (_____) = _____

energy (_____) = _____

[You need to remember these equations.]

When charge flows through a resistor, the energy transferred makes the resistor _____ .

When charge flows around a circuit, the _____ supplied by the battery is equal to

the _____ transferred to all the components in the circuit.

B Use the idea of electrons to explain why a wire gets hot when a current flows.

_____

_____

_____

C a You might sometimes have felt a small shock from a car door handle. When this happens a current of about 4 mA flows for about 0.1 s.

Calculate the charge that flows.

charge = _____ C

b When you use a toaster, the wire inside the toaster heats up. The toaster is connected to the mains at 230 V, and a charge of 300 000 C flows.

Calculate the energy transferred.

energy transferred = _____ J

D A student calculates the energy transferred to the components in a circuit as 120 J, and the energy transferred by the battery as 125 J.

Explain why these two values are not exactly equal.

_____

_____

_____

P5.5 Appliances and efficiency

A Fill in the gaps to complete the sentences.

A domestic meter measures how much _____ is transferred by appliances in your home.

You can calculate energy supplied to an appliance, or the useful energy, using the equations:

Energy supplied (_____) = _____

Useful energy (_____) = _____

Useful power (_____) = _____

[Include units. You need to be able to remember these equations.]

B Jules uses an electric oven to cook a chicken. The oven needs a potential difference of 230 V and has a current flowing through it of 15 A.

a Calculate the power of the oven. Write your answer to an appropriate number of significant figures.

power = _____ W

b Jules now wants to calculate the energy transferred to the oven.

Name the other quantity that he will need. _____

C a It takes 30 minutes to cook a pizza in an oven with a power of 2000 W.

Calculate the energy transferred.

energy transferred = _____ J

b Another oven is 70% efficient. It supplies 4000 kJ.

Calculate the useful energy.

useful energy = _____ J

c Both ovens supply the same amount of useful energy. Calculate the efficiency of the oven in part **a**.

efficiency = _____ %

P5 Practice questions

01 **Table 1** shows some information about three wires in a plug.

Table 1

Wire	Colour	Statement
A		connected to the metal casing of an appliance
B		at 230 V
C		at 0 V

01.1 Complete the table by writing the colour of each wire. [2 marks]

01.2 Write down the letter of the earth wire.

_____ [1 mark]

01.3 Write down the letters of the **two** wires that make a complete circuit with an appliance.

_____ [1 mark]

01.4 Write down the name of the network of wires and transformers to which the plug is connected.

_____ [1 mark]

02 Complete **Table 2** by writing down for each material the part of a plug it is used for and why.

Table 2

Material	The part of a plug it is used for	Reason
hard plastic		
flexible plastic		
copper		

03 A student has some hair straighteners. The label says: 1100 W, 230 V. Calculate the fuse that she needs to use in the plug. The fuses available are 3 A, 5 A, and 13 A. [4 marks]

_____

_____

04 A student does a survey of the appliances in her kitchen. **Table 3** shows the results. All the appliances work on a potential difference of 230 V (the mains).

Table 3

Appliance	Power rating in W	Current in A
kettle	1200	5.2
microwave	800	3.5
refrigerator	420	1.8

04.1 Calculate the energy transferred when you microwave popcorn for 2 minutes. [3 marks]

_____

_____

04.2 Calculate the charge that flows through a kettle during the 6 minutes that it takes to boil. [3 marks]

_____

_____

04.3 The refrigerator is on all day and all night. Calculate the energy transferred in one day. [3 marks]

_____

_____

05 A student uses an oscilloscope to displace a low-voltage alternating p.d. The screen is shown in **Figure 1**.

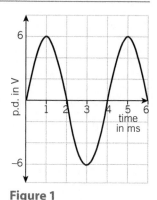

Figure 1

05.1 Calculate the frequency of the p.d. [4 marks]

_____

_____

05.2 Give **two** features of the graph that tell you it is not showing mains electricity. [2 marks]

_____

_____

06 The graph in **Figure 2** shows how the efficiency of two types of washer has changed over time.

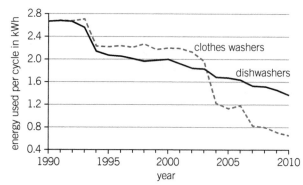

Figure 2

The energy needed to clean a load in a dishwasher is 3 MJ. Compare the energy wasted by a typical dishwasher in 1990 with that of a typical dishwasher in 2010. [6 marks]

_____

_____

_____

_____

_____

P5 Checklist

	Student Book	☺	☺	☹
I can write down what direct current is and what alternating current is.	5.1			
I can describe what is meant by the live wire and the neutral wire of a mains circuit.	5.1			
I can describe the National Grid.	5.1			
I can describe how to use an oscilloscope to measure the frequency and peak potential difference of an alternating current.	5.1			
I can describe what the casing of a mains plug or socket is made of and explain why.	5.2			
I can write down what is in a mains cable.	5.2			
I can write down the colours of the live, neutral, and earth wires.	5.2			
I can explain why a three-pin plug includes an earth pin.	5.2			
I can describe how power and energy are related.	5.3			
I can use the power rating of an appliance to calculate the energy transferred in a given time.	5.3			
I can calculate the electrical power supplied to a device from its current and potential difference.	5.3			
I can work out the correct fuse to use in an appliance.	5.3			
I can calculate the flow of electric charge given the current and time.	5.4			
I can write down the energy transfers when electric charge flows through a resistor.	5.4			
I can describe how the energy transferred by a flow of electric charge is related to potential difference.	5.4			
I can link the electrical energy supplied by the battery in a circuit to the energy transferred to the electrical components.	5.4			
I can calculate the energy supplied to an electrical appliance from its current, its potential difference, and how long it is used for.	5.5			
I can work out the useful energy output of an electrical appliance.	5.5			
I can work out the output power of an electrical appliance.	5.5			
I can compare different appliances that do the same job.	5.5			

P6.1 Density

A Fill in the gaps to complete the sentences.

Density depends on _____ and _____, and is measured in _____.

You can calculate density using the equation:

$$\text{density (_____)} = \frac{\rule{4cm}{0.4pt}}{\rule{4cm}{0.4pt}}$$

[Include units. You need to be able to remember this equation]

From this you can work out that mass = _____ × _____, and volume = $\dfrac{\rule{3cm}{0.4pt}}{\rule{2cm}{0.4pt}}$.

You use a digital balance to measure _____, and a ruler or measuring cylinder to measure

_____ .

An object will float on water if its density is _____ _____ that of water.

B Calculate the **density** of a person with a mass of 55.0 kg and a volume of 0.0700 m³.

density = _____ kg/m³`

C A ship is much heavier than a small pebble. But whilst the ship can float on water, a pebble sinks. Explain why.

_____

_____

_____

_____

D Compare how you would find the density of a stone cube and how you would find the density of an irregular piece of modelling clay.

_____

_____

_____

_____

E Material **X** has a density of 1.5 g/cm³. Material **Y** has a density of 5 g/cm³. A block of material **Y** has the same mass as 10 cm³ of material **X**.

Calculate the volume of the block of material **Y**.

_____

P6.2 States of matter

A Fill in the gaps to complete the sentences.

In a _____ the particles move about randomly and are far apart. In a _____ the particles move at random and are in contact with each other. In a _____ the particles are held next to each other in fixed positions.

A _____ is the least energetic state of matter, and a _____ is the most energetic state of matter.

When a substance changes state, the _____ stays the same because the number of particles stays the same.

B Use the particle arrangements of solids and liquids to explain why the densities of a solid and liquid metal are similar.

_____

_____

_____

C a Label each of the arrows in the diagram with the correct name of the change of state.

b Explain why liquid is a more energetic state than a solid.

_____

_____

D The water in a cat's bowl evaporates over time.

a Write down what happens to the number of molecules of water in the bowl over time.

_____

b Explain how mass is conserved in this situation.

_____

_____

c Explain why this is a **physical change**.

_____

P6.3 Changes of state

A Fill in the gaps to complete the sentences.

The melting point of a pure substance is the temperature at which it _____ or _____ , and the boiling

point is the temperature at which it _____ or _____ .

You can find the melting point or boiling point from the _____ section of a temperature–time graph.

_____ occurs throughout a liquid at its boiling point, but _____ occurs from the surface of a liquid at a
temperature below its boiling point.

B Tick the boxes to show whether each statement is true for boiling or evaporation. You may need to tick **both**
boxes for some statements.

Statement	✓ if true for boiling	✓ if true for evaporation
This process happens at the **boiling point** of the liquid.		
The mass does not change.		
The particles escape only from the surface of the liquid.		
This process happens at or below the boiling point of the liquid.		

C Jo took some ice out of the freezer and put it on a plate in a warm room. This graph shows what happened to the
temperature of the ice over time.

temperature

time

a Label the diagram using the words below.

        melting point   **liquid**   **solid + liquid**   **solid**

b Jo then reversed the process by putting the melted ice back in
the freezer.

Sketch a graph of temperature against time for this process.

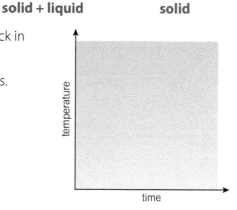

c Suggest **one** change you would need to make when labelling this graph compared with the graph in
activity **C**.

_____

41

P6.4 Internal energy

A Fill in the gaps to complete the sentences.

If the temperature of a substance increases, its internal energy _____ .

The strength of the forces of _____ between the particles explains why a substance is a solid, liquid, or gas.

If you heat a substance and its temperature rises, the _____ energy of its particles increases. If you

heat a substance and its temperature *does not* rise, the _____ energy of its particles increases.

The pressure of a gas on a surface is caused by the particles of the gas repeatedly _____ the surface.

B Tick **all** the descriptions of energy that apply to the **internal energy** of a substance in different states.

Energy	✓ if included in internal energy
the kinetic energy of the particles in a gas	
the energy of vibration of the particles in a solid	
the gravitational potential energy of the particles in a liquid	
the kinetic energy of a whole solid	

C Write a sentence to explain why a gas exerts a pressure on the surfaces that enclose it.

_____

_____

D a Compare a liquid getting hotter and a solid cooling down. Use the idea of changes in the potential energy and kinetic energy of particles.

_____

_____

_____

b Explain why you can heat a substance but not observe a temperature rise. Use the ideas of internal, kinetic, and potential energy.

_____

_____

_____

_____

_____

P6.5 Specific latent heat

A Fill in the gaps to complete the sentences.

Latent heat is the _____ you need to transfer to a substance to change its state without changing its

_____ .

Specific latent heat of fusion or vaporisation is the energy you need to transfer to a substance to melt or boil a mass of

_____ of the substance without changing its _____ .

You can calculate the energy transferred using this equation, with units in the brackets:

thermal energy for a change of state (_____) = _____

You can measure the specific latent heat of ice or water using a low-voltage heater to _____ the ice, or

to _____ the water.

B Compare latent heat and specific latent heat, using their definitions and units.

_____

_____

_____

C A student wanted to measure the thermal energy needed to melt ice. He measured the mass of some ice. He melted the ice and measured the mass of the water. The **specific latent heat of fusion** of ice is 334 kJ/kg, and the mass of water melted was 0.03 kg.

Calculate the thermal energy for this change of state. Give your answer to an appropriate number of significant figures.

thermal energy for melting ice = _____ J

D For a given substance, the specific latent heat of vaporisation is higher than the specific latent heat of fusion. Explain why. Use the ideas of potential energy, kinetic energy, and bonds.

_____

_____

_____

_____

_____

P6.6 Gas pressure and temperature

A Fill in the gaps to complete the sentences.

Gas molecules colliding with the surfaces in contact with the gas cause gas _____ .

If the temperature of a gas in a sealed container increases, the pressure _____ because:

- there are _____ impacts per second
- the force of the impacts _____ .

You can see evidence for the _____ motion of gas molecules by observing smoke particles.

B Write a sentence to describe a piece of evidence for the motion of gas molecules.

_____

_____

_____

_____

C The pressure of a gas increases as it is heated. Explain why, by describing its particles.

_____

_____

D Molly carried out an investigation into the relationship between temperature and pressure. Here is the graph that she plotted from her results.

a Describe **two** variables that Molly needs to keep constant. Explain why.

_____

_____

_____

_____

b Molly heats the gas in a steel container. As the temperature increases, the seal of the container leaks. Suggest what this would do to Molly's graph.

_____

_____

_____

_____

P6 Practice questions

01 Write the definition of latent heat. [1 mark]
_____
_____

0.2 Compare each of the following changes of state, in terms of particles.

02.1 Freezing and melting. [2 marks]
_____
_____

02.2 Evaporation and boiling. [2 marks]
_____
_____

03 You have 0.5 m³ cubes of three types of solid material: **A, B,** and **C. Table 1** shows the masses of the cubes.

Table 1

Material	Mass in kg
A	2000
B	3700
C	3000

03.1 Calculate the density of the material with the lowest density. [3 marks]
_____
_____

03.2 The density of water is 1000 kg/m³. Explain what would happen to the cubes if you put them in water. [2 marks]
_____
_____

04 **Figure 1** shows a graph of the temperature of a liquid as it is heated.

Figure 1

04.1 Write down the boiling point of the liquid.
_____ [1 mark]

04.2 Write down the time when the particles have the least potential energy. _____ [1 mark]

04.3 Write down the time when the particles have the least kinetic energy. _____ [1 mark]

04.4 Describe what happens to the internal energy of the liquid between 5 and 8 minutes. [2 marks]
_____
_____

05.1 Sort these steps into the correct order to describe how to measure the specific latent heat of fusion of ice. Write the letters in order below. [4 marks]

A Look at the joulemeter to record the energy transferred by the heater during the 10 minutes.

B Turn the heater on.

C Allow the ice to melt for 10 minutes and measure the mass of water collected.

D Put ice in a funnel, put the funnel in a beaker on a digital balance, and put a heater into the ice, but do not turn it on.

E Use the equation specific latent heat = energy/mass of water collected, to find the specific latent heat.

Correct order: _____
_____

05.2 Suggest and explain why a student using this method might observe a value for the latent heat of fusion that is higher than the accepted value. [3 marks]
_____
_____
_____

06 An ice cube of mass 2 g melts, and the water heats up to room temperature, which is 20 °C.

The specific latent heat of melting of ice is 334 000 J/kg, and the specific heat capacity of water is 4200 J/kg °C.

Calculate the total energy supplied to the ice cube. Use these equations:

thermal energy for a change of state = mass × specific latent heat

change in thermal energy = mass × specific heat capacity × temperature change [4 marks]
_____
_____
_____
_____
_____
_____
_____
_____
_____

P6 Checklist

	Student Book	☺	☺	☹
I can define density and write down its unit.	6.1			
I can describe how to measure the density of a solid object or a liquid.	6.1			
I can use the density equation to calculate the mass or the volume of an object or a sample.	6.1			
I can describe how to tell from its density if an object will float in water.	6.1			
I can describe the different properties of solids, liquids, and gases.	6.2			
I can describe the arrangement of particles in a solid, a liquid, and a gas.	6.2			
I can explain why gases are less dense than solids and liquids.	6.2			
I can explain why the mass of a substance that changes state stays the same.	6.2			
I can write down what the melting point and the boiling point of a substance mean.	6.3			
I can describe what you need to do to melt a solid or to boil a liquid.	6.3			
I can explain the difference between boiling and evaporation.	6.3			
I can use a temperature–time graph to find the melting point or the boiling point of a substance.	6.3			
I can describe how increasing the temperature of a substance affects its internal energy.	6.4			
I can explain the different properties of a solid, a liquid, and a gas.	6.4			
I can describe how the energy of the particles of a substance changes when it is heated.	6.4			
I can explain in terms of particles why a gas exerts pressure.	6.4			
I can write down what latent heat means as a substance changes its state.	6.5			
I can write down what specific latent heat of fusion and of vaporisation mean.	6.5			
I can use specific latent heat in calculations.	6.5			
I can describe how to measure the specific latent heat of ice and of water.	6.5			
I can describe how a gas exerts pressure on a surface.	6.6			
I can describe how changing the temperature of a gas in a sealed container affects the pressure of the gas.	6.6			
I can explain why raising the temperature of a gas in a sealed container increases the pressure of the gas.	6.6			
I can describe how to see evidence of gas molecules moving around at random.	6.6			

P7.1 Atoms and radiation

A Fill in the gaps to complete the sentences.

A radioactive substance contains _____ nuclei that usually become _____ after emitting radiation.

Radioactive sources emit three main types of radiation: _____ , _____ , and

_____ .

You cannot predict when a nucleus will emit radiation, so we say radioactive decay is _____ .

B Give an example, apart from radioactive decay, of a random process, and explain why it is random.

_____

_____

C Complete the table by writing the missing types of radiation, the symbol for each type, and the part of the atom that emits the radiation. You may need to use the same word more than once.

Type of radiation	Symbol	Part of the atom that emits the radiation
alpha		
	γ	
	β	

D A scientist is investigating a radioactive rock specimen using a Geiger counter. There is a reading on the counter that shows the number of counts per second.

a Describe what is causing the reading to change.

_____

_____

b Describe and explain how the scientist could work out whether the specimen is emitting alpha radiation.

_____

_____

_____

c Explain in terms of atoms why the Geiger counter produces a reading with some rocks but not others.

_____

_____

_____

P7.2 The discovery of the nucleus

A Fill in the gaps to complete the sentences.

Rutherford used _____ particles to probe atoms. He fired them at a thin metal foil and discovered

that most of them went through, but some were scattered by _____ angles.

He could not explain this scattering using the _____ _____ model.

Rutherford's model said that most of the mass of the atom is in a _____ ,

_____ charged nucleus in the centre of an atom.

B Look at the diagram of Rutherford's experiment with alpha particles.

a In the right-hand diagram, two of the alpha particles are scattered back in the direction that they came from. Draw the paths of those two alpha particles.

b Describe and explain the link between the number of alpha particles scattered backwards and Rutherford's model of the atom.

_____

_____

_____

C Compare Rutherford's model of the atom with Bohr's model of the atom.

_____

_____

_____

_____

_____

P7.3 Changes in the nucleus

A Fill in the gaps to complete the sentences.

Isotopes of an element are atoms with the _____ number of protons but a

_____ number of neutrons. They have the _____ atomic number but

_____ mass numbers.

When a nucleus emits an alpha particle it loses _____ protons and _____

neutrons. The mass number goes down by _____, and the atomic number goes down by

_____ .

When a nucleus emits a beta particle a _____ changes to a _____ and

emits an _____ . The mass number _____ _____

_____, and the atomic number goes up by _____ .

B **a** Here are some symbols for atoms. The names of the elements have been replaced with an '**X**'.

Circle the **two** isotopes of the same element.

$$^{14}_{6}\text{X} \qquad ^{14}_{7}\text{X} \qquad ^{12}_{6}\text{X}$$

b Explain the decision you made in part **a**.

_____

_____

C Compare alpha decay and beta decay by describing the changes in the nucleus.

_____

_____

_____

_____

D Use the periodic table to write balanced equations for the decay of the following isotopes by alpha and beta decay.

The alpha decay of $^{226}_{88}\text{Ra}$: $^{226}_{88}\text{Ra} \rightarrow$ $\qquad$ $+$

The beta decay of $^{218}_{84}\text{Po}$: $^{218}_{84}\text{Po} \rightarrow$ $\qquad$ $+$

E Compare γ emission and neutron emission.

_____

_____

_____

_____

_____

_____

A Fill in the gaps to complete the sentences.

Alpha radiation is stopped by _____, beta radiation is stopped by _____, and the intensity of gamma radiation is reduced by _____.

In air, alpha radiation has a range of ____ ____ _____, beta radiation has a range of _____ _____, and gamma radiation has an _____ range.

An alpha particle consists of _____ protons and _____ neutrons, a beta particle is a _____-moving _____, and gamma radiation is _____ radiation.

Alpha radiation is the _____ ionising, and gamma radiation is the _____ ionising.

All three types of radiation _____ substances as they pass through them, which can _____ or kill living cells.

B A student has a sample that is radioactive. She wants to identify the type or types of radiation emitted by the sample.

Describe and explain how the student can use a Geiger counter and samples of different materials to identify the emitted radiation.

_____

_____

_____

_____

C **a** In the table, write the correct type of radiation (α, β, γ) next to each description.

it has an infinite range in air		it consists of a fast moving electron	
it is moderately ionising		it is the most ionising	
it has a range of about 1 m in air		it has a range of a few cm in air	
it consists of two protons and two neutrons		it consists of electromagnetic radiation	

b One description is missing. Write the description and the type of radiation here:

_____

D A friend finds out that a pack of strawberries has been irradiated. He thinks that eating them will give him cancer.

Suggest what you can say to reassure him.

_____

_____

_____

P7.5 Activity and half-life

A Fill in the gaps to complete the sentences.

The half-life of a radioactive isotope is the average time it takes for the number of nuclei of the isotope to

_____ .

The count rate of a Geiger counter decreases as the activity of a radioactive source _____ .

In one half-life the activity and the number of atoms of a radioactive isotope will _____ .

You can find the count rate of a radioactive isotope after n half-lives by dividing the initial count rate by

_____ .

B Explain the difference between **count rate** and **activity**.

_____

_____

C Tick **all** of the correct definitions of **half-life**.

Definition	✓ if correct
Half-life is the time for the number of alpha particles to halve.	
Half-life is the time for the number of unstable nuclei to halve.	
Half-life is the time for the activity to halve.	
Half-life is the time for the amount of radiation to halve.	

D Look at the graph. It shows the decay of radioactive iodine.

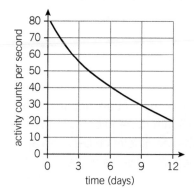

a Write down the half-life of the iodine. _____

b Calculate the activity after three half-lives.

_____

_____

c Calculate the time when the activity becomes five counts per second. Show how you calculated the time.

_____

_____

d Calculate the ratio of net decline of the iodine after six half-lives.

_____

_____

P7 Practice questions

01 A radioactive sample emits alpha, beta, and gamma radiation.

01.1 Compare these three types of radiation, by describing the types of particle or wave, types of absorber, and ranges in air. [3 marks]

_____

_____

_____

_____

01.2 Explain the link between the ionising power of the different types of radiation and their range in air. [3 marks]

_____

_____

02 Ideas about the atom have changed over time. Here are three descriptions of models of atoms.

A a positively charged mass with negatively charged electrons embedded in it

B a positively charged nucleus with electrons in specific energy levels around it

C a positively charged nucleus with electrons in orbit around it

Write down the **observations** that led to the changes in the model of the atom over time. Use the descriptions above. [6 marks]

_____

_____

_____

_____

_____

_____

03 Here are some statements about what happens to the atomic mass and atomic number of a nucleus that decays.
Write a possible type of decay (α, β, γ) for each statement.

03.1 The mass number goes down by four. [1 mark]

_____

03.2 The atomic number does not change. [1 mark]

_____

03.3 The atomic number goes up by one. [1 mark]

_____

03.4 The mass number does not change. [1 mark]

_____

04 **Figure 1** is a graph of the activity of a radioactive sample of californium against time.

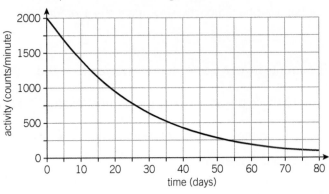

Figure 1

04.1 The activity of californium halves every 18 days. Write down, in terms of nuclei, what else halves in 18 days. [1 mark]

_____

_____

04.2 Calculate the activity of californium after five half-lives. [4 marks]

_____

_____

_____

05 A radiographer in a hospital is preparing a sample of technetium-99. She shields herself from the radiation that the technetium emits.

05.1 Explain why radiation is hazardous to the human body. [1 mark]

_____

_____

05.2 A different isotope of technetium, technetium-101, has a half-life of 15 minutes and emits beta radiation. Describe the difference between a nucleus of technetium-99 and a nucleus of technetium-101. [2 marks]

_____

_____

05.3 Complete the nuclear equation for the beta decay of technetium-101. [2 marks]

$$^{101}_{43}\text{Tc} \rightarrow \boxed{}\text{Ru} + ^{0}_{-1}\beta$$

05.4 The doctor injects a patient with technetium-99. She uses the gamma radiation detected with a special camera to diagnose kidney problems in the patient. Explain why the doctor does not use technetium-101 for this procedure. [3 marks]

_____

_____

P7 Checklist

	Student Book	☺	😐	☹
I can write down what a radioactive substance is.	7.1			
I can write down the types of radiation given out from a radioactive substance.	7.1			
I can write down what happens when a radioactive source emits radiation (radioactive decay).	7.1			
I can write down the different types of radiation emitted by radioactive sources.	7.1			
I can describe how the nuclear model of the atom was established.	7.2			
I can explain why the 'plum pudding' model of the atom was rejected.	7.2			
I can describe what conclusions were made about the atom from experimental evidence.	7.2			
I can explain why the nuclear model was accepted.	7.2			
I can write down what an isotope is.	7.3			
I can describe how the nucleus of an atom changes when it emits an alpha particle or a beta particle.	7.3			
I can represent the emission of an alpha particle from a nucleus.	7.3			
I can represent the emission of a beta particle from a nucleus.	7.3			
I can write down how far each type of radiation can travel in air.	7.4			
I can describe how different materials absorb alpha, beta, and gamma radiation.	7.4			
I can describe the ionising power of alpha, beta, and gamma radiation.	7.4			
I can explain why alpha, beta, and gamma radiation are dangerous.	7.4			
I can write down what the half-life of a radioactive source means.	7.5			
I can write down what the count rate from a radioactive source means.	7.5			
I can describe what happens to the count rate from a radioactive isotope as it decays.	7.5			
I can calculate the count rate after a given number of half-lives.	7.5			

P8.1 Vectors and scalars

A Fill in the gaps to complete the sentences.

Displacement is the _____ in a given direction.

A vector quantity is a physical quantity that has _____ and _____ . A scalar quantity is a physical quantity that has _____ only.

You can represent a vector quantity with an arrow. The direction of the arrow tells you the _____ of the vector, and the length of the arrow tells you the _____ of the vector.

B **a** Write down a scalar quantity that you have measured in an experiment, and explain why it is not a vector.

_____

_____

  b Explain why vectors are shown by arrows.

_____

C A tortoise walks 20 cm west, then 30 cm north, and then 40 cm east. You may want to draw a diagram to help your calculations.

Calculate:

a the final displacement of the tortoise

_____

b the total distance that the tortoise travels.

_____

D A toy boat is travelling north across a lake. The driving force of the engine on the boat is 20 N. The force of the wind on the boat is 5 N west. The wind is blowing at 90° to the direction of motion of the boat.

Draw a scale diagram to show the forces acting on the boat.

Write down the scale that you used.

The scale used is: _____ cm = _____ N

P8.2 Forces between objects

A Fill in the gaps to complete the sentences.

Forces can change the _____ of an object, change the _____ of an object, or start a

_____ object moving. Forces are measured in _____ .

A _____ force is a force that acts on objects only when they touch each other.

When two objects interact they always exert _____ and _____ forces on each other.

B Write down **one** example of each of the following effects of a force.

a A force can change the shape of an object. Example:

_____

_____

b A force can change the motion of an object. Example:

_____

_____

c A force can start an object moving that was at rest. Example:

_____

_____

C Compare **contact forces** with **non-contact forces**. Give **one** example of each type of force.

_____

_____

_____

D **a** Write down **Newton's third law**.

_____

_____

b Explain why the **two** forces acting on you as you sit on a chair writing an answer to this question are **not** an example of Newton's third law.

_____

_____

_____

c Describe how Newton's third law explains how you walk across the floor.

_____

_____

d Describe how Newton's third law explains how a space rocket takes off when fuel in the rocket is ignited.

_____

_____

_____

P8.3 Resultant forces

A Fill in the gaps to complete the sentences.

The resultant force is a single force that has the _____ effect as all the forces acting on an object.

An object stays at rest or moves with a steady speed when the resultant force on it is _____ .

The speed or direction of an object changes when the resultant force on it is _____ than

_____ .

If there are two forces acting on an object along the same line you _____ them if they act in the same

direction, and find the _____ if they act in opposite directions.

A _____ force diagram of an object shows the forces acting on it.

B Explain how the resultant force on a moving cyclist could be zero.

_____

_____

C a Here are some descriptions of the motion of objects. Circle the descriptions where the resultant force is zero.

stationary   **speeding up**   **slowing down**   **moving at a steady speed**

b The table shows some pairs of forces that act along the same line.

Complete the table by finding the resultant force of each pair of forces.

Write the magnitude and the direction (left or right).

Force to the left in N	Force to the right in N	Resultant force magnitude in N	Resultant force direction (left or right?)
7	3		
10	20		
80	150		

D a You are floating on the surface of a swimming pool. Draw arrows as seen from the side to show the forces acting on you, and the free-body diagram of you.

The force of the water on you	The force of the Earth on you	A free-body diagram of you
●	●	●

b Explain how you know that the forces on you are balanced. How did you show this in the free-body diagram?

_____

_____

_____

P8.4 Centre of mass

A Fill in the gaps to complete the sentences.

The centre of mass of an object is the _____ where all the mass is concentrated.

For a uniform object such as a sphere or a cube, the centre of mass is in the _____ of the object.

A suspended object will stop swinging when the centre of mass is _____ the point of suspension.

The centre of mass of a symmetrical object is along its _____ of symmetry.

B Explain in terms of centre of mass why the height of a racing car is much smaller than the height of a family car.

_____

_____

C Put a dot in each object below where you expect its centre of mass to be.

D High winds can be a problem for lorries. In the diagrams below, the centre of mass of the lorry and the pivot point are labelled.

a Add an arrow to each diagram to show the force of the Earth (weight) on the lorry.

b Use the diagrams to explain why a very strong wind is needed to topple a lorry.

_____

_____

_____

_____

P8.5 The parallelogram of forces

A Fill in the gaps to complete the sentences.

The parallelogram of forces is the _____ diagram of two force vectors.

The parallelogram of forces is used to find the _____ of two forces that do not act along the same line.

To make a parallelogram of forces diagram, you need a ruler, a pencil, a blank sheet of paper, and a _____ .

The resultant is the diagonal of the parallelogram that starts at the _____ of the two forces.

B Describe the type of situation where you need to use a parallelogram of forces, and the type of situation where you do not.

_____

_____

C A force of 60 N and a force of 40 N act on the same object. The angle between the forces is 30°.

a Draw a scale diagram to represent the forces on the object.

b Use the diagram to find the resultant force.

resultant force = _____ N

D Two tug boats pull a tanker along a river.

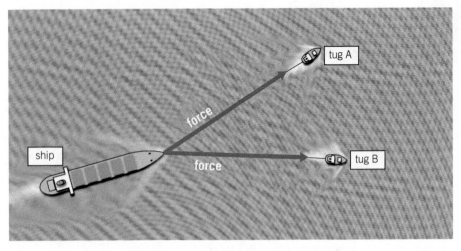

a Draw the resultant force on the tanker.

b In each of the situations below, describe the effect on the resultant force.

i Tug B pulls twice as hard as tug A.

_____

_____

ii The rope attached to tug A breaks.

_____

_____

P8.6 Resolution of forces

A Fill in the gaps to complete the sentences.

When you are resolving a force, you are finding two components that are at _____ to each other that have a resultant force that is equal to that force.

You can resolve a force in two perpendicular directions by drawing a _____ with adjacent sides along the two directions so that the _____ represents the force vector.

The resultant force is zero for an object in _____.

An object at _____ is in equilibrium because the resultant force on it is zero.

B **a** You are sitting on a hillside. Write down the angle between the slope and the normal component of your weight. _____

b Compare the magnitude of the normal force when you are sitting on flat ground with the normal force when you are sitting on the hillside.

_____

_____

C The arrow shows a resultant force of 10 N. Use a scale diagram to calculate the horizontal and vertical components of the force that is equal to this resultant force.

D A student puts a box on a ramp. He then lifts the end of the ramp until the box is just about to move.

a Draw and label the forces that are acting on the box along the slope and normal to the slope. Two of these forces are components.

b Give the two measurements that the student needs to make to measure the frictional force on the box, and explain why.

_____

_____

c The student replaces the box with a trolley. Explain the effect on the height the student needs to lift the ramp to make the trolley move.

_____

_____

P8 Practice questions

01 Explain why force is a vector but temperature is **not** a vector. [1 mark]

_____

_____

02 Circle the contact forces in the list below. [2 marks]

gravity   **friction**   **tension**

magnetic attraction   **air resistance**

03 **Figure 1** shows some forces acting on a box.

5N   2N   5N   5N   5N   2N

←— A —→   ←— B —→   ←— C —→
                                        2N

resultant:_____   resultant:_____   resultant:_____

Figure 1

03.1 Draw an arrow (where appropriate) to show the resultant force acting on each box. Label with the magnitude of the force. [3 marks]

03.2 Explain which of the boxes is in equilibrium. [2 marks]

_____

_____

03.3 **Table 1** describes what each box **A**, **B**, and **C** in **03.1** is doing before the forces are applied to them.

Table 1

Box	What the box is doing	What the box will start to do
A	not moving	
B	moving at 3 m/s to the left	
C	moving at 3 m/s to the left	

Complete the table to describe what each box will start to do when the forces shown in **03.1** are applied to them. [3 marks]

04.1 Describe a journey where the magnitude of the displacement is the **same** as the magnitude of the distance travelled. [1 mark]

_____

_____

04.2 Describe a journey where the magnitude of the displacement is **different** from the magnitude of the distance travelled. [1 mark]

_____

05 You are lying on a sofa and then get up and stand on the floor.

Compare the magnitudes of the reaction forces acting on you when you are lying on the sofa and when you are standing on the floor. [1 mark]

_____

06 A child sits on a swing made from an old tyre (**Figure 2**).

06.1 Write down what you know about the position of the centre of mass of the child and swing. [1 mark]

_____

_____

Figure 2

06.2 A friend of the child pushes the tyre to the left so that the angle between the rope and the vertical is 30°. The weight of the child and tyre is 50 N.

Use a free-body scale diagram to find the tension in the rope. [4 marks]

07 A zip wire is strung between two trees, and people hold on to a pulley as they race across the wire.

One person is in the centre of the wire.

Figure 3

The angle between each half of the wire and the horizontal is 10°. The tension in each half of the wire is 2000 N.

07.1 Draw a parallelogram of forces to scale. Use it to calculate the weight of the person. [4 marks]

07.2 Use the parallelogram of forces diagram to explain why it is preferable that the angle should not be very small. [3 marks]

_____

_____

_____

P8 Checklist

	Student Book	☺	😐	☹
I can write down what displacement is.	8.1			
I can write down what a vector quantity is.	8.1			
I can write down what a scalar quantity is.	8.1			
I can describe how to represent a vector quantity.	8.1			
I can write down what forces can do.	8.2			
I can write down the unit of force.	8.2			
I can write down what a contact force is.	8.2			
I can describe the forces being exerted when two objects interact.	8.2			
I can describe what a resultant force is.	8.3			
I can describe what happens if the resultant force on an object is zero.	8.3			
I can describe what happens if the resultant force on an object is greater than zero.	8.3			
I can calculate the resultant force when an object is acted on by two forces acting along the same line.	8.3			
I can write down what the centre of mass of an object is.	8.4			
I can write down where the centre of mass of a metre ruler is.	8.4			
I can find the centre of mass of an object suspended from a fixed point.	8.4			
I can find the centre of mass of a symmetrical object.	8.4			
I can write down what a parallelogram of forces is.	8.5			
I can write down what a parallelogram of forces is used for.	8.5			
I can write down what is needed to draw a scale diagram of a parallelogram of forces.	8.5			
I can use a parallelogram of forces to find the resultant of two forces.	8.5			
I can describe what resolving a force means.	8.6			
I can describe how to resolve a force into two components.	8.6			
I can define equilibrium.	8.6			
I can explain why an object at rest is in equilibrium.	8.6			

A Fill in the gaps to complete the sentences.

You can calculate speed using this equation:

speed (_____) = ────────────────────

[You need to remember this equation.]

The distance–time graph for a _____ object is a horizontal straight line, and the distance–time graph for

an object moving with a _____ _____ is a straight line that slopes upwards.

The _____ of a distance-time graph for an object tells you the object's speed.

B a An ice skater glides across the ice and travels a distance of 11 m in 2.7 s.
   Calculate his speed.

   speed = _____

b Give an assumption that you need to make in order to calculate this speed.

   _____

C A skateboarder travels at 5.5 m/s.

a Calculate how far he travels in 15 s.

   distance travelled = _____ m

b Calculate the time it takes him to travel 70 m.

   time = _____ s

D Use the data on the graph to describe the motion of the object in section **A** of the
graph and compare it with the motion in section **B**.

_____

_____

_____

_____

_____

_____

P9.2 Velocity and acceleration

A Fill in the gaps to complete the sentences.

Velocity is the speed in a given _____ .

A _____ quantity is a physical quantity that has magnitude and direction. A _____ quantity is a physical quantity that has magnitude only.

You can calculate acceleration using this equation:

$$\text{acceleration} \ (\underline{\quad}) = \frac{\overline{\rule{6cm}{0pt}}}{\rule{5cm}{0pt}}$$

[You need to remember this equation.]

Deceleration is the change of velocity per second when an object _____ _____ .

An object moving in a circle with a constant speed has a _____ that is constantly changing.

B **a** Explain the difference between speed and velocity.

_____

    b Circle the vectors in the list below.

10 m/s north   **25 mph**   **38 mph south**   **10 m/s**   **38 mph**

C **a** A car travels in a circle of radius 25 m in 20 seconds.

       Calculate its speed.

       $\pi = 3.14$

       speed = _____ m/s

    b Explain why the car is accelerating.

_____

_____

    c Compare the speed and velocity of the car when it is at opposite sides of the circle.

_____

_____

D **a** A plane takes off. Its velocity changes from rest to a speed of 100 m/s over half a minute.

       Calculate the acceleration of the plane. Give the unit.

       acceleration _____

    b The plane prepares to land. Its velocity changes from 155 m/s to 140 m/s over 2 minutes.

       Calculate the acceleration of the plane. Give the unit.

       acceleration = _____

P9.3 More about velocity–time graphs

A Fill in the gaps to complete the sentences.

You can use a _____ _____ linked to a computer to measure velocity changes.

The _____ of the line on a velocity-time graph tells you the acceleration. If the line is horizontal the

acceleration is _____ .

If the object has a positive acceleration the line has a _____ _____ .

If the object is decelerating the line has a _____ _____ .

The area under the line on a velocity–time graph tells you the _____ _____ .

B Describe **one** advantage and **one** disadvantage of using a motion sensor to produce a velocity–time graph.

_____

_____

C Complete the table to explain how to interpret a velocity–time graph.

If the line on the graph is...	...then the object is...
...straight, and going up	
...straight, and going down	
...straight, and horizontal	

D Here is the velocity–time graph for a train travelling between two stations.

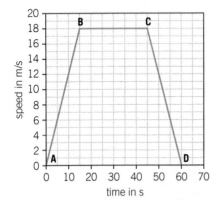

a Calculate the acceleration between points **A** and **B**.

b Calculate the total distance travelled.

c Explain how you can write down the acceleration between points **C** and **D** without doing a calculation.

_____

_____

_____

P9.4 Analysing motion graphs

A Fill in the gaps to complete the sentences.

You can find the speed from a distance–time graph by finding the _____ of the line on the graph.

You can find the acceleration from a velocity–time graph by finding the _____ of the line on the graph.

You can find the distance travelled from a velocity–time graph by finding the _____ _____ the line on the graph.

If the speed is changing, you can find the speed at any instant in time from a distance–time graph by finding the

_____ of the _____ to the line on the graph.

B Compare a horizontal line on a distance–time graph and a horizontal line on a speed–time graph.

_____

_____

_____

C An object travels a distance of 20 m in 10 s at a steady speed, and is then stationary for 10 s. Then its speed steadily increases for 10 s.

Describe the gradient of a velocity–time graph for the motion by completing the table below.

During...	The gradient of the line on the graph will be...
0–10 seconds	
10–20 seconds	
20–30 seconds	

D A student produces distance–time graphs for two objects, **X** and **Y**.

a Calculate the speed of object **X** at 10 s.

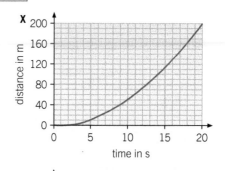

b Calculate the speed of object **Y** at 10 s.

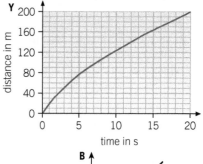

c Here are the two velocity–time graphs for objects **X** and **Y**. Suggest and explain which graph describes the motion of object **X**.

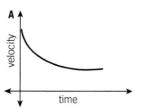
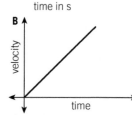

_____

_____

P9 Practice questions

01 Complete **Table 1** by calculating speed, distance, or time. [3 marks]

Table 1

Steady speed in m/s	Distance in m	Time in s
	52	7.0
12		0.02
115	3000	

02 Riya drops her phone. It accelerates from a vertical speed of 0 m/s to a speed of 0.5 m/s in a time of 0.05 s. Calculate the acceleration. [3 marks]

_____

_____

_____

03 **Figure 1** is a graph showing a person's journey.

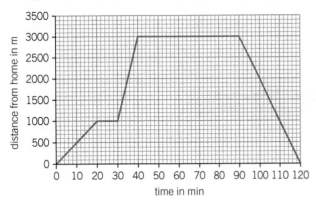

Figure 1

03.1 Write down **one** of the distances from home at which the person was stationary. [1 mark]

_____

03.2 Calculate the fastest speed shown on the graph. [2 marks]

_____

_____

03.3 Suggest how the measurements were obtained for this graph. [2 marks]

_____

_____

04 **Figure 2** shows the velocity–time graph for a cyclist.

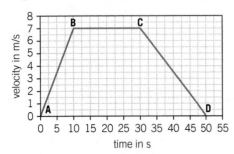

Figure 2

04.1 Use the graph to write down or calculate the acceleration of the cyclist between points **A**, **B**, **C**, and **D**. [5 marks]

_____

_____

_____

_____

_____

_____

04.2 A student uses a velocity of 7 m/s, a time of 50 s, and the equation distance = speed × time to calculate that the distance travelled by the cyclist is 350 m.
Explain why this method is **incorrect**. [1 mark]

_____

_____

04.3 Calculate the **correct** distance travelled by the cyclist. [2 marks]

_____

_____

04.4 The cyclist travels in a large circle at a steady speed of 5 m/s. Suggest and explain one other measurement that you need to make to calculate the acceleration of the cyclist. [3 marks]

_____

_____

_____

05 **Figure 3** shows a distance–time graph for a drag racer.

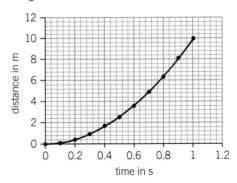

Figure 3

Use the graph to calculate the speed of the racer after 0.5 s. [3 marks]

_____

_____

_____

_____

P9 Checklist

	Student Book	☺	😐	☹
I can calculate speed is calculated for an object moving at constant speed.	9.1			
I can use a distance–time graph to determine whether an object is stationary or moving at constant speed.	9.1			
I can write down what the gradient of the line on a distance–time graph can tell you.	9.1			
I can use the equation for constant speed to calculate distance moved or time taken.	9.1			
I can write down the difference between speed and velocity.	9.2			
I can calculate the acceleration of an object.	9.2			
I can write down the difference between acceleration and deceleration.	9.2			
I can explain that motion in a circle involves constant speed but changing velocity.	9.2			
I can measure velocity change.	9.3			
I can write down what a horizontal line on a velocity–time graph tells you.	9.3			
I can use a velocity–time graph to work out whether an object is accelerating or decelerating.	9.3			
I can write down what the area under a velocity–time graph tells you.	9.3			
I can calculate speed from a distance–time graph where the speed is constant.	9.4			
I can calculate speed from a distance–time graph where the speed is changing.	9.4			
I can calculate acceleration from a velocity–time graph.	9.4			
I can calculate distance from a velocity–time graph.	9.4			

P10.1 Forces and acceleration

A Fill in the gaps to complete the sentences.

If the resultant force on an object increases, the acceleration will _____ , as long as the _____ stays the same.

If two objects of different masses have the same resultant force acting on them, the acceleration of the object with the bigger mass will be _____ .

You can calculate the resultant force acting on an object using this equation:

resultant force (_____) = _____

[You need to remember this equation.]

The inertia of an object is its tendency to stay at _____ or in _____ motion.

B Describe how the acceleration of an object depends on the force and the mass.

_____

_____

_____

_____

C Write a sentence to describe **inertial mass**.

_____

_____

D a Calculate the resultant force on a ball that has a mass of 105 g and is accelerating at 3.20 m/s². Include the units.

resultant force = _____

b You exert the same resultant force on another ball that has a mass of 210 g.

Calculate the acceleration of this ball.

acceleration = _____ m/s²

c Explain whether the ball in part **a** or the ball in part **b** has the greater inertia.

_____

_____

d Explain what you would need to do to the resultant force to accelerate the 210 g ball at 3.20 m/s².

_____

_____

P10.2 Weight and terminal velocity

A Fill in the gaps to complete the sentences.

The weight of an object is the _____ acting on the object due to gravity. The mass is the quantity of

_____ in the object.

If there is only gravity acting on an object, then it will accelerate at about _____ m/s^2.

When an object falls, it eventually reaches _____ velocity. This happens when the weight equals the

_____ force on the object. At this velocity, the resultant force on the object is _____ .

B a A person reads their 'weight' on some bathroom scales. The measurement is 75 kg. Explain in terms of physics
why this is wrong.

_____

_____

b Complete the table with the correct values of mass and weight.

Mass in kg	Gravitational field strength in N/kg	Weight in N
0.2	10	
	27	30
0.035	1.6	

C a The only force acting on an object is gravity.

Write down the acceleration of the object.

acceleration = _____ m/s^2

b Explain why a stone dropped into a deep lake will not have the same acceleration.

_____

_____

c The stone is dropped from a point 1 m above the surface of the lake.

Describe and explain what happens to the velocity of the stone from the point it is dropped until it hits the
bottom of the lake.

_____

_____

_____

_____

d Sketch a graph to show how the speed of the stone changes with time. Mark an 'x' on the part of the graph
where the resultant force is zero.

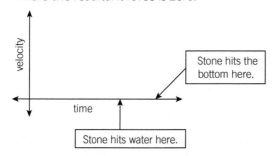

velocity

time

Stone hits the bottom here.

Stone hits water here.

P10.3 Forces and braking

A Fill in the gaps to complete the sentences.

Friction and air resistance oppose the _____ force of a vehicle.

The stopping distance of a vehicle depends on the _____ distance and the _____ distance.

Poor weather conditions, poor vehicle maintenance, and speed affect the _____ distance. Poor reaction

time and high speed both affect the _____ distance.

You can work out the braking force of a vehicle using this equation:

force (_____) = _____

B a Complete the table by defining each distance and stating **three** factors that affect each one. One has been done for you.

Distance	Definition	Depends on
braking distance		1. speed 2. 3.
thinking distance		1. speed 2. 3.

b Describe a situation in which a lorry might have the same stopping distance as a car.

_____

_____

C a Calculate the force of the brakes on a car of mass 1000 kg that has a deceleration of 4 m/s². Include the unit.

force = _____

b The car is initially travelling at 30 m/s. It needs to stop in a distance of 120 m to avoid an accident. Explain whether the car will avoid the accident.

_____

_____

c Suggest why the braking distance of a car is not proportional to the speed.

_____

_____

P10.4 Momentum

A Fill in the gaps to complete the sentences.

You can calculate momentum using this equation:

momentum (_____) = _____

[You need to remember this equation.]

A closed system is a system in which the total momentum before an event is _____ _____ _____ the total momentum after the event. This is called _____ of momentum.

B A lorry has twice the mass of a car.

Explain how both the lorry and the car can have the same momentum.

_____

_____

_____

C A plane of mass 300 000 kg is taxiing on a runway at 2 m/s.

Calculate the momentum of the plane. Include the unit.

momentum = _____

D A student investigates two colliding trolleys. Motion sensors measure the speed of the trolleys. Before the collision, trolley **X** (mass 300 g) moves with a velocity of 2 m/s to the left, and collides with trolley **Y** (mass 150 g) moving to the right. After the collision, both trolleys stop.

a Calculate the velocity of trolley **Y** before the collision.

velocity of trolley **Y** before the collision = _____ m/s

b Write an assumption that you made when doing this calculation.

_____

c In a different collision, the two trolleys stick together after the collision and move together to the left. Compare the motion of the trolleys before the collision in this experiment with their motion in part **a**.

_____

_____

P10.5 Forces and elasticity

A Fill in the gaps to complete the sentences.

An object is _____ if it returns to its original shape after you remove the force that you have used to deform it.

The extension of an object is the _____ between the length when you stretch it and its original length.

The extension of a spring is _____ _____ to the force applied to it. This is only true if you do

not go beyond the limit of _____ . This is a _____ relationship.

If you *do* go beyond the limit of _____ , then the relationship becomes _____–_____ , and

the force and extension are no longer _____ .

B a Describe and explain what you could do to show that the material used to make plastic bags shows plastic, and not elastic, behaviour.

_____

_____

  b Name the **two** quantities that are no longer proportional if the limit of proportionality is exceeded.

_____

C a A student stretches a spring with a spring constant of 40 N/m. The spring is 3.0 cm long at the start, and 4.5 cm long after the student applies a force.

Calculate the force applied.

  force = _____ N

  b Calculate the length of the spring if the force on the spring is trebled.

  length = _____ cm

D This graph shows the extension of a sample as a student increases the force applied to the sample.

a Explain how you know that the sample obeys Hooke's law.

_____

b Describe what the student would observe if she hung weights of up to 6 N on the sample and then removed the weights.

_____

_____

c Calculate the spring constant of the sample.

  spring constant = _____ N/m

P10 Practice questions

01 Explain how and why speed affects the stopping distance of a car. [6 marks]

_____

_____

_____

_____

_____

_____

02.1 Complete **Table 1** by calculating the resultant force, mass, or acceleration. [3 marks]

Table 1

Mass	Acceleration in m/s²	Force
1.3 g	2	
1.5 kg		1 kN
	7×10^{-3}	2 N

02.2 Explain how you know that the mass in **Table 1** is the inertial mass. [1 mark]

_____

_____

03 A student drops a light ball. One of the forces acting on the ball is the force of the Earth on it.

03.1 Write down another name for this force. [1 mark]

_____

03.2 Write down the other force acting on the ball as it falls. [1 mark]

_____

03.3 **Table 2** shows the speed of the ball as it falls.

Table 2

Time in s	Speed in m/s
0.0	0
0.1	1
0.2	2
0.3	3
0.4	3
0.5	3

Write down the value of the terminal velocity.

_____ [1 mark]

03.4 Describe and explain the motion of the ball between 0.3 s and 0.5 s. [3 marks]

_____

_____

_____

04 An aircraft of mass 250 000 kg lands on an aircraft carrier. There is a hook at the bottom of the aircraft that latches onto thick wires, which stretch to

bring the aircraft to a stop. The aircraft lands at a speed of 60 m/s and comes to a stop over a distance of 50 m.

04.1 Calculate the deceleration of the aircraft. [3 marks]

_____

_____

04.2 Calculate the force of the wire on the aircraft. [3 marks]

_____

_____

04.3 The wire extends by 10 m.

Calculate the spring constant of the wire. [2 marks]

_____

_____

04.4 Calculate the momentum of the aircraft as it latches onto the wire. [2 marks]

_____

_____

05 Two students investigate the effect of changing the mass of a trolley and hanger system on its acceleration, as shown in **Figure 1**. In both experiments, the mass of the trolley is 300 g.

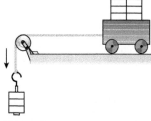

Figure 1

Student **A** simply adds weights to the hanger. Student **B** adds weights to the hanger by taking them from the trolley. **Figure 2** shows the graphs of their results.

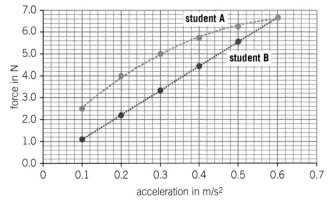

Figure 2

Explain the shapes of the graphs. [6 marks]

_____

_____

_____

_____

_____

_____

_____

P10 Checklist

	Student Book	☺	☺	☹
I can describe how the acceleration of an object depends on the size of the resultant force acting upon it.	10.1			
I can describe the effect that the mass of an object has on its acceleration.	10.1			
I can describe how to calculate the resultant force on an object from its acceleration and its mass.	10.1			
I can write down what the inertia of an object means.	10.1			
I can describe the difference between mass and weight.	10.2			
I can describe and explain the motion of a falling object acted on only by gravity.	10.2			
I can write down what terminal velocity means.	10.2			
I can write down what can be said about the resultant force acting on an object that is falling at terminal velocity.	10.2			
I can describe the forces that oppose the driving force of a vehicle.	10.3			
I can write down what the stopping distance of a vehicle depends on.	10.3			
I can write down what can increase the stopping distance of a vehicle.	10.3			
I can describe how to estimate the braking force of a vehicle.	10.3			
I can calculate momentum.	10.4			
I can write down the unit of momentum.	10.4			
I can describe what momentum means in a closed system.	10.4			
I can describe what happens when two objects push each other apart.	10.4			
I can write down what 'elastic' means.	10.5			
I can describe how to measure the extension of an object when it is stretched.	10.5			
I can describe how the extension of a spring changes with the force applied to it.	10.5			
I can write down what the limit of proportionality of a spring means.	10.5			

P11.1 The nature of waves

A Fill in the gaps to complete the sentences.

You can use waves to transfer _____ and energy.

A wave that oscillates perpendicular (at 90°) to the direction of energy transfer is called a

_____ wave. Examples of these waves are the _____ on the

surface of water, and _____ waves, such as light.

A wave that oscillates parallel to the direction of energy transfer is called a _____ wave.

An example of this kind of wave is a _____ wave produced by a loudspeaker.

Mechanical waves need a _____ to travel through.

B A student attaches a rope to a door handle and pulls it tight.

a Describe how she can make a **transverse** wave on the rope.

_____

_____

b Explain why she cannot make a **longitudinal** wave on the rope.

_____

_____

c Write down what is transferred by **both** transverse and longitudinal waves.

_____

C Here are some diagrams of seismic (earthquake) waves. The arrow shows the direction of motion of the waves in the ground.

a Explain which wave is transverse.

_____

_____

b The diagram of the longitudinal wave shows compressions and rarefactions. Explain the difference between a compression and a rarefaction.

_____

_____

c Explain why seismic waves are mechanical.

_____

_____

P11.2 The properties of waves

A Fill in the gaps to complete the sentences.

The amplitude of a wave is the _____ displacement of a point on the wave from its undisturbed

position. This could be the height of a wave _____, or the height of a wave _____.

The wavelength of a wave is the distance from a point on a wave to the _____ point on the next

wave. This could be from one wave _____ to the next wave _____.

You can calculate the period and speed of a wave using these equations:

$$\text{period} (_____) = \frac{\rule{5cm}{0pt}}{\rule{5cm}{0pt}}$$

speed (_____) = _____

[**You need to remember these equations.**]

B A musical instrument makes a note with a frequency of 200 Hz.

Calculate the period of the sound wave. Give the unit.

period = _____

C **a**   A sound wave has a wavelength of 0.50 m and a frequency of 660 Hz.

Calculate the speed of the wave. Give the unit.

speed = _____

b   Write down the wavelength of the wave if the frequency is doubled. _____

c   A seismic wave has a speed of 4 km/s and a wavelength of 70 m.

Calculate the frequency.

frequency = _____ Hz

D A student writes some sentences about the properties of electromagnetic waves.

> *The distance from one peak to the next peak is the wavelength. Frequency is the number of waves per second. The unit of frequency is seconds. The unit of wavelength is metres. The distance from a peak to a trough is the amplitude.*

a   Underline the **correct** sentences.

b   Rewrite the **incorrect** sentences.

_____

_____

_____

_____

_____

_____

P11.3 Reflection and refraction

A Fill in the gaps to complete the sentences.

Plane waves are reflected from a straight barrier at the _____ angle as the incident waves. This is because their _____ and wavelength do not change.

Plane waves crossing a boundary between two different materials are _____ unless they cross the barrier at _____ incidence.

At the boundary of two different materials, _____ occurs, because the speed and wavelength of the waves change at the boundary.

At a boundary between two different materials, waves can be _____ or _____.

B **a** Complete the diagrams to show what happens when waves in a ripple tank are reflected (diagram **X**), and when they are refracted (diagram **Y**).

reflection   refraction

X   Y

b On diagram **X** label the angle of incidence and the angle of refraction.

c Describe and explain what happens to the wavelength of the waves that are refracted.

_____

_____

d Here is another diagram of waves crossing a boundary in a ripple tank.

Compare the waves in diagram **Y** and the waves in diagram **Z**. Explain whether or not diagram **Z** shows refraction.

Z

_____

_____

_____

_____

C Suggest why you can see the bottom of a swimming pool, but the bottom of a deep ocean is completely dark.

_____

_____

P11.4 More about waves

A Fill in the gaps to complete the sentences.

You hear an echo when a sound wave _____ from a smooth, hard surface.

You can measure the speed of sound by measuring the _____ interval between seeing a short loud sound being made and hearing it. If you also measure the _____ you can use the equation:

$$\text{speed} \, (\underline{\quad}) = \frac{\rule{6cm}{0.4pt}}{\rule{6cm}{0.4pt}}$$

to find the speed of sound.

B a Define an echo.

_____

b Describe how to use two wooden blocks and a large building to measure the speed of sound.

_____

_____

_____

_____

_____

_____

_____

c A student stands a distance of 150 m from a wall and measures a time delay of 0.90 s between making a sound and hearing the echo.

Calculate the speed of sound.

Give your answer to the appropriate number of significant figures.

speed = _____ m/s

d Estimate the uncertainty in:

● the student's measurement of the distance. _____

● the student's measurement of the time. _____

e Use your estimates to describe the effect of these two uncertainties on the value of speed that you calculated in part **c**.

_____

_____

_____

_____

P11 Practice questions

01 Look at the diagram in **Figure 1**.

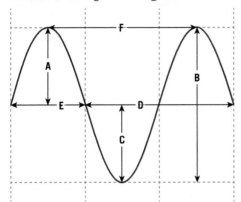

Figure 1

01.1 Complete **Table 1** by writing the correct letters from **Figure 1** that describe each wave property. There are **two** letters you do not need to use. [4 marks]

Table 1

Wave property	Letters
wavelength	_____ or _____
amplitude	_____ or _____

01.2 The wave in **Figure 1** is moving from left to right. Explain how you know that this is a transverse wave. [1 mark]

02.1 Describe the difference between mechanical waves and electromagnetic waves. [1 mark]

02.2 Compare the speed of mechanical waves with the speed of electromagnetic waves. [1 mark]

03 A sound wave underwater has a frequency of 1000 Hz and a wavelength of 1.5 m.

03.1 Calculate the speed of sound in water. [3 marks]

03.2 Calculate the period of the wave. [3 marks]

04 A student makes plane waves in a ripple tank. She puts a barrier in the tank to turn the waves through an angle of 90°.

04.1 Draw the incoming waves and the barrier position that would achieve this. [2 marks]

04.2 Write what is happening to the wave. [1 mark]

04.3 Describe **two** other things that can happen to waves when they interact with matter that do not involve a change in direction. [2 marks]

05 A group of students collect some data from an experiment to measure the speed of sound. They stand a long way from a wall and make a sound. They measure the distance as 200 m and the time as 1.3 s.

05.1 Describe how they measured the time in this experiment. [2 marks]

05.2 Use the data to calculate the speed of sound. Give your answer to 2 significant figures. [4 marks]

05.3 Write down **one** improvement that the students could make to their experiment. [1 mark]

06 A student wants to investigate how the depth of water in a ripple tank affects the speed of plane waves.

06.1 Name the **independent** variable in this investigation. _____ [1 mark]

06.2 Name the **dependent** variable in this investigation. _____ [1 mark]

06.3 Sketch and explain the graph that the student might obtain from this investigation. Label the axes. [3 marks]

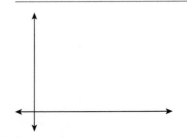

P11 Checklist

	Student Book	☺	☺	☹
I can describe what waves can be used for.	11.1			
I can describe what transverse waves are.	11.1			
I can write down what longitudinal waves are.	11.1			
I can write down which types of wave are transverse and which are longitudinal.	11.1			
I can define the amplitude, frequency, and wavelength of a wave mean.	11.2			
I can describe how the period of a wave is related to its frequency.	11.2			
I can write down the relationship between the speed, wavelength, and frequency of a wave.	11.2			
I can use the wave speed equation in calculations.	11.2			
I can draw the patterns of reflection and refraction of plane waves in a ripple tank.	11.3			
I can determine whether plane waves that cross a boundary between two different materials are refracted.	11.3			
I can explain reflection and refraction using the behaviour of waves.	11.3			
I can describe what can happen to a wave when it crosses a boundary between two different materials.	11.3			
I can write down what sound waves are.	11.4			
I can write down what echoes are.	11.4			
I can describe how to measure the speed of sound in air.	11.4			

P12.1 The electromagnetic spectrum

A Fill in the gaps to complete the sentences.

The waves of the electromagnetic spectrum are:

radio, _____ , _____ , _____ light, _____ , _____ - _____ , and gamma rays.

This list of waves is in order from the _____ to the _____ wavelength, and from the _____ to the

_____ frequency and energy.

The human eye can only detect _____ light, which has a range of wavelengths from _____ nm to

_____ nm.

Electromagnetic waves transfer energy from a _____ to an _____ .

You can use the wave equation:

_____ = _____

to calculate the frequency and wavelength of electromagnetic waves.

[*You need to remember this equation.*]

B Choose a correct electromagnetic wave to complete the sentences below. There may be more than one correct answer.

a _____ have a wavelength longer than microwaves.

b Ultraviolet waves transfer more energy than _____ .

c X-rays transfer less energy than _____ .

d The frequency of visible light is higher than the frequency of _____ .

e The speed of infrared waves is the same as the speed of _____ .

f _____ can have a wavelength that is bigger than 1 km.

C The wavelength of microwaves is 3 cm, and their frequency is 10 000 000 000 Hz.

Calculate the speed of microwaves. Include the unit.

wave speed = _____

D Calculate the smallest and the largest frequencies of waves that can be detected by the human eye. Use the wave speed that you calculated in activity **C**.

$1 \text{ nm} = 1 \times 10^{-9} \text{m}$

smallest frequency: _____   largest frequency: _____

P12.2 Light, infrared, microwaves, and radio waves

A Fill in the gaps to complete the sentences.

White light contains all the _____ of the visible spectrum.

Infrared radiation is used to carry _____ from remote control handsets and inside _____ fibres.

Your mobile phone uses _____ waves and _____ . Satellite TV uses _____ , and radio and TV broadcasting uses _____ _____ .

Different types of electromagnetic radiation are hazardous in different ways. _____ waves and _____ can heat parts of the body, and _____ radiation can cause skin burns.

B Compare how infrared radiation is used in infrared cameras and in remote controls.

_____

_____

_____

_____

C **a** Describe how the human body detects infrared waves and light waves.

_____

b Remote controls use infrared waves. Describe how you can use a remote control to show that infrared waves can be reflected.

_____

_____

c Write down how you know that the walls of your house transmit microwaves and radio waves.

_____

_____

D The frequency of the microwaves used in a microwave oven is 2.45 GHz. The frequency of the microwaves used in a mobile phone is 0.1 GHz.

a Calculate the ratio of the frequency of waves used in the microwave oven to the frequency of waves used in the mobile phone.

ratio of the frequencies = _____

b The energy of an electromagnetic wave is proportional to its frequency. So the ratio of the frequencies is the same as the ratio of the energies.

Comment on the ratio that you calculated in part **a**, by describing how the microwaves in the microwave oven and the microwaves in the mobile phone are used.

_____

_____

_____

_____

_____

P12.3 Communications

A Fill in the gaps to complete the sentences.

We use radio waves of different frequencies for different purposes. This is because the wavelength and frequency

affect the _____ they travel, how much they _____ out, and how much

_____ they carry. Carrier waves are waves that are used to carry _____. They do this

by varying their _____.

We use _____ to transmit satellite television signals.

We need further _____ before we will know whether or not mobile phones are safe to use.

We send signals by _____ or _____ radiation down thin transparent fibres called optical fibres.

B a Sort the uses of radio waves and microwaves in order of those that need the shortest wavelength to those that need the longest wavelength.

Write the letters in the correct order below.

W local radio stations **Y** international radio stations

X satellite TV and satellite phones **Z** TV broadcasting

Correct order: _____

b Use the ideas of range, spreading out, and absorption to explain why:

i microwaves instead of radio waves are used for satellite communication.

_____

_____

ii national and local radio stations use radio waves of different wavelengths.

_____

_____

C a Describe how infrared waves are used for communication.

_____

b Explain why using infrared waves in this way is safer than using microwaves.

_____

D The diagram shows how a radio signal is produced from a sound (audio) signal.

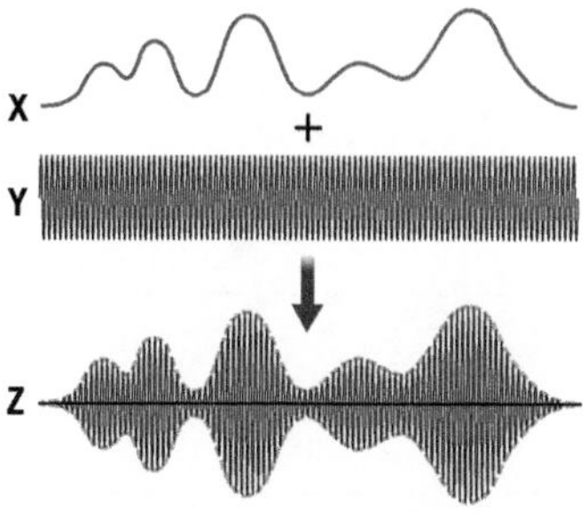

a Write the correct letter next to each type of signal:

Carrier wave		Audio signal		Radio signal	

b Use the waves shown in the diagram to compare how a radio transmitter and a radio receiver work.

_____

_____

_____

_____

P12.4 Ultraviolet waves, X-rays, and gamma rays

A Fill in the gaps to complete the sentences.

Ultraviolet waves have a _____ wavelength than visible light, and can harm the _____
and eyes.

X-rays are used by doctors in _____ to make X-ray images.

Gamma rays are used to kill harmful _____ in food, to _____ surgical

equipment, and to kill _____ cells.

When _____ radiation travels through matter, it can make uncharged atoms charged.

X-rays and gamma rays can cause _____ to living tissue when they pass through it.

B a Explain why ultraviolet waves can cause skin cancer, but not cancers inside the body.

_____

_____

b Explain how X-rays or gamma rays can cause cancer.

_____

_____

_____

_____

_____

_____

_____

c Describe the link between the wavelength of the radiation and the damage that it can do to DNA.

_____

_____

d Describe **one** way that damaging cells using ionising radiation is put to good use.

_____

_____

C Explain how the ozone layer acts like sunscreen.

_____

_____

D Compare X-rays and gamma rays by describing their wave properties, dangers, and method of production.

_____

_____

_____

_____

_____

P12.5 X-rays in medicine

A Fill in the gaps to complete the sentences.

X-rays are used in hospitals to make X-ray images and to destroy _____ cells at or near the body surface.

X-rays are _____ radiation, so they can damage living tissue when they pass through it.

_____-energy X-rays are used for imaging, and _____-energy X-rays are used to kill cancer cells.

Bones and teeth _____ more X-rays passing through the body than soft tissue does.

B Describe **one** use in hospitals of:

a low-energy X-rays

_____

b high-energy X-rays

_____

c gamma rays.

_____

C The diagram shows an X-ray being taken of a broken bone.

a Suggest why the photographic film is in a light-proof wrapper.

_____

b Complete the diagram to show the paths of the X-rays.

Explain how the paths you have drawn produce the image on the photographic film.

_____

_____

_____

_____

X-ray beam

skin

bone

limb
table

film in light-proof wrapper

c Suggest and explain which metal should be used to cover the parts of the body not being X-rayed.

_____

D Doctors use X-rays to diagnose medical problems. Each X-ray scan increases a person's annual dose of radiation.

The table shows radiation doses for two types of X-ray scan, and a typical person's average annual radiation dose.

	Dose in millisieverts
Chest X-ray scan	0.1
Whole-body X-ray scan	10
Average annual dose	2

a Calculate the number of chest X-rays that would give a patient the equivalent of their average annual dose of radiation.

_____

_____

b Suggest how the data in the table might influence whether or not a doctor decides to X-ray a patient.

_____

_____

_____

_____

P12 Practice questions

01 There are seven types of wave in the electromagnetic spectrum.

01.1 Write the waves in decreasing order of wavelength. [6 marks]

_____

_____

01.2 Describe the order of the waves in part **01.1** in terms of frequency. [1 mark]

_____

01.3 Write the wave that the human eye can detect.

_____ [1 mark]

02.1 Name **two** electrical devices used in the home that **emit** electromagnetic radiation. Next to each device, state the type of radiation that it emits. [4 marks]

1. _____

2. _____

02.2 Name **two** electrical devices used in the home that **absorb** electromagnetic radiation. Next to each device, state the type of radiation that it absorbs. [4 marks]

1. _____

2. _____

03 Radio waves and microwaves are used in communications.

03.1 The radio waves or microwaves are carrier waves. Describe what a carrier wave is. [1 mark]

_____

_____

03.2 Write down one advantage of using short wavelength radio waves instead of long wavelength radio waves. [1 mark]

_____

_____

_____

03.3 Explain how you know from their uses that microwaves have a range of wavelengths. [3 marks]

_____

_____

_____

03.4 Explain why it is difficult to do experiments to work out the size of the risk to human health of using mobile phones. [2 marks]

_____

_____

04 Look at the X-ray image in **Figure 1**, which was produced in a hospital.

04.1 This radiation is ionising. Describe what 'ionising' means. [1 mark]

_____

_____

Figure 1

04.2 Explain why some parts of the image are black, and some are white. [3 marks]

_____

_____

04.3 Describe **two** precautions that the person making this image should take to reduce the risk of injury. Explain why the risk is reduced. [3 marks]

_____

_____

05 The frequency of a radio wave is 100 000 000 Hz, and it travels at 300 000 000 m/s. Calculate the wavelength of the radio wave. [3 marks]

_____

06 Doctors can use gamma rays to produce images of internal organs, such as the kidneys as shown in **Figure 2**.

signal to computer

gamma camera

γ -rays

patient

Figure 2

06.1 Explain why you cannot use X-rays to produce an image of the kidneys. [1 mark]

_____

_____

06.2 Suggest why the precautions that are taken to reduce the risk of injury to a patient having an X-ray cannot be used for this patient. [2 marks]

_____

_____

P12 Checklist

	Student Book	☺	☻	☹
I can write down the parts of the electromagnetic spectrum.	12.1			
I can write down the range of wavelengths within the electromagnetic spectrum that the human eye can detect.	12.1			
I can describe how energy is transferred by electromagnetic waves.	12.1			
I can calculate the frequency or wavelength of electromagnetic waves.	12.1			
I can describe the nature of white light.	12.2			
I can list some uses of infrared radiation, microwaves, and radio waves.	12.2			
I can write down what mobile phone radiation is.	12.2			
I can explain why these types of electromagnetic radiation are hazardous.	12.2			
I can explain why radio waves of different frequencies are used for different purposes.	12.3			
I can write down which waves are used for satellite TV.	12.3			
I can describe how to decide whether or not mobile phones are safe to use.	12.3			
I can describe how optical fibres are used in communications.	12.3			
I can describe what a carrier wave is.	12.3			
I can describe the differences between ultraviolet and visible light.	12.4			
I can list some uses of X-rays and gamma rays.	12.4			
I can write down what ionising radiation is.	12.4			
I can explain why ultraviolet waves, X-rays, and gamma rays are dangerous.	12.4			
I can describe what X-rays are used for in hospitals.	12.5			
I can explain why X-rays are dangerous.	12.5			
I can write down which parts absorb X-rays when they pass through the body.	12.5			
I can explain the difference between the uses of low- and high-energy X-rays in hospitals.	12.5			

P13.1 Magnetic fields

A Fill in the gaps to complete the sentences.

When you bring two magnet poles together, they will _____ if the poles are the same, and

_____ if the poles are different.

The magnetic field lines of a bar magnet are in a direction from the _____ pole of the magnet to

the _____ pole of the magnet.

If you put a piece of magnetic material in a magnetic field, it will become an _____ magnet.

Permanent magnets are made out of _____ because it does not lose its magnetism easily.

A magnetic material such as _____ does lose its magnetism easily.

B Explain how you can use a permanent magnet to work out whether a bar of metal is magnetic or a permanent magnet.

_____

_____

C When you lift a paperclip out of a bowl of paperclips that have been used in magnetism experiments, it can attract other paperclips.

a Explain why new paperclips made of the same material do not do this.

_____

_____

b Suggest and explain the material that the paperclips are made of.

_____

_____

_____

D **a** Sketch the magnetic field lines around this bar magnet. Add arrows to the magnetic field lines to show the direction of the magnetic field.

N	S

b Explain what happens to the needle of a compass that is placed close to the magnet then moved a very long distance from the magnet.

_____

_____

_____

_____

P13.2 Magnetic fields of electric currents

A Fill in the gaps to complete the sentences.

To draw the magnetic field pattern around a current-carrying wire, you need to draw _____ centred on the wire.

The magnetic field lines inside a solenoid are _____ and all in the same direction.

You can increase the strength of the magnetic field around a wire or in a solenoid by _____ the current. If you reverse the current, you _____ the direction of the magnetic field.

For a uniform magnetic field, the field lines are all _____ to each other.

B a Complete the diagram to show the shape and direction of the magnetic field around a current-carrying wire.
The current in the wire is in a direction into the paper.

b Describe how you would need to change your diagram if the current was reversed.
_____

C a Describe the difference between a solenoid and an electromagnet.
_____
_____

b Describe the difference between the magnetic field inside an electromagnet and the magnetic field outside the electromagnet.
_____
_____

D A student investigates how the current in an electromagnet affects its strength.

a Name **two** variables that the student will need to control in this experiment.
_____

b Sketch a graph of the mass of iron filings that the electromagnet picks up against current. Label the axes.

c Explain the shape of the graph.
_____
_____

d Explain **one** advantage and **one** disadvantage of using iron filings rather than paperclips in this experiment.
_____
_____
_____

P13.3 The motor effect

A Fill in the gaps to complete the sentences.

In the motor effect, increasing the current, the strength of the magnetic field, or the length of the conductor

will _____ the force. Reversing the direction of the current or the magnetic field will

_____ the direction of the force.

An electric motor has a _____ that turns when a current passes through it.

Magnetic _____ _____ is a measure of the strength of a magnetic field.

You can use the equation, with units in the brackets:

force (___) = _____

to calculate the force on a current-carrying conductor at right angles to the lines of a magnetic field.

B A 0.2 m long wire carrying a current of 2.0 A is placed in a magnetic field where the flux density is 0.4 T.

a Describe how you need to place a wire in the magnetic field to produce the maximum force on the wire.

_____

b Calculate the force on the wire.

force = _____ N

c Calculate the current that would need to flow in the wire to produce a force of 0.05 N on the wire.

current = _____ A

C Look at the diagram. It shows a long wire carrying a current down through the centre of a piece of card. The direction of this current is into the paper.

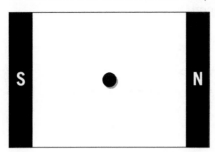

a Use Fleming's left-hand rule to determine the force on the wire. Draw an arrow on the diagram to show the direction of the force.

b Motors use this effect. Describe a simple electric motor.

_____

_____

c Complete the table by summarising the effect of making changes to a motor.

If you...	...then the motor will...
reverse the direction of the current in the coil	
decrease the current in the coil	
increase the strength of the magnets	

P13 Practice questions

01 Write down the rule for working out what happens when you bring the poles of two magnets together. [2 marks]

_____

_____

02 Some magnets are permanent, but you can make an electromagnet by winding wire around a core.

02.1 Explain why refrigerator magnets are made with permanent magnets and not electromagnets. [2 marks]

_____

_____

_____

_____

02.2 Describe and explain what would happen if you used steel and not iron for the core of an electromagnet. [2 marks]

_____

_____

03 A student wants to investigate the magnetic field around a current-carrying wire. He uses the equipment shown in **Figure 1**.

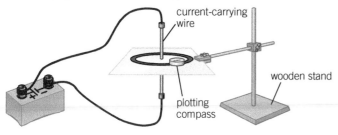

Figure 1

03.1 Write down **one** hazard in doing this experiment, and how to reduce the risk of injury. [2 marks]

_____

_____

03.2 **Figure 2** shows the magnetic field patterns around two current-carrying wires.

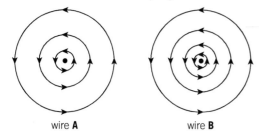

wire **A**   wire **B**

Figure 2

Use **Figure 2** to write down and explain the difference between the currents flowing in the wires. [2 marks]

_____

_____

_____

04.1 Table 1 shows some different solenoids. Put them in order of strength starting with the strongest, by writing the letters in order below.

Table 1

Solenoid	Number of turns	Current in A	Core
A	40	3	air
B	40	3	iron
C	30	2	air
D	20	2	air

Correct order: (strongest) ___, ___, ___, ___ (weakest) [2 marks]

04.2 Explain why solenoid **B** is stronger than solenoid **A**. [3 marks]

_____

_____

05 **Figure 3** shows a simple electric motor. The motor is spinning clockwise.

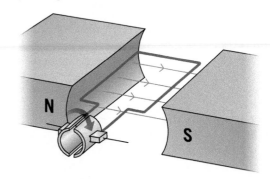

Figure 3

05.1 Draw two arrows on **Figure 3** to show the current flow in the long sides of the coil. [2 marks]

05.2 Write down what would happen to the motor if the current was reversed. [1 mark]

_____

05.3 Each side of the coil is 5 cm long, has a force on it of 0.005 N, and carries a current of 1.2 A.

Calculate the magnetic flux density. Give the unit. [3 marks]

magnetic flux density = _____

P13 Checklist

	Student Book	☺	😐	☹
I can write down the force rule for two magnetic poles near each other.	13.1			
I can describe the pattern of magnetic field lines around a bar magnet.	13.1			
I can write down what induced magnetism is.	13.1			
I can explain why steel, not iron, is used to make permanent magnets.	13.1			
I can describe the pattern of the magnetic field around a straight wire carrying a current.	13.2			
I can describe the pattern of the magnetic field in and around a solenoid.	13.2			
I can describe how the strength and direction of the magnetic field varies with position and current.	13.2			
I can write down what a uniform magnetic field is.	13.2			
I can write down what an electromagnet is.	13.2			
I can write down how to change the size and direction of the force on a current-carrying wire in a magnetic field.	13.3			
I can describe how a simple motor works.	13.3			
I can write down what magnetic flux density means.	13.3			
I can calculate the force on a current-carrying wire at right angles to the lines of a magnetic field.	13.3			

Answers

P1.1

A store

heating, waves, electric current, a force

gravitational potential, kinetic

kinetic, surroundings, heating

B examples:

   a use a battery-powered appliance, e.g., a torch

   b lift an object, e.g., a book, onto a shelf

   c pull back / stretch an elastic band / spring

C at the beginning, there is energy in the chemical store of the food; this is transferred to the chemical store of your muscles when you eat; when you turn the handle, the energy in the chemical store decreases, and the energy in the thermal store of the surroundings increases; the transfer is made by forces when you turn the handle; by an electric current when you generate electricity; by heating and radiation when the wire inside the light bulb gets hot and emits light

P1.2

A created, destroyed, conservation, all

closed, energy

change

B it would swing forever and would always reach the same height; if there is no friction then the system is closed; there is no energy transfer in or out of the pendulum

C a the ball would bounce forever; it would always reach the same height;

   b the child would swing forever; the child would always reach the same height; the bungee jumper would bounce forever;

   c the bungee jumper would extend the bungee cord, bounce back up to the same height that she jumped from, and repeat

D a the system of the roller coaster itself is not closed, because of friction.

   b the system of the roller coaster, its surroundings, and its track is closed.

P1.3

A force, transferring

J, force (N) × distance moved in the direction of the force (m)

air resistance, friction, heat

B work done (J) = force (N) × distance (m)

$$= 15\,(N) \times 180\,(m)$$
$$= 2700\,J$$

C 20 J, 3 J, 125 kJ, 1 kJ

D independent variable: mass of sand

dependent variable: force on newton-meter

control variables: type of surface, tub, distance moved

the student needs to measure distance moved, mass of sand, and force needed; she needs to calculate the work done using the equation work = force × distance

she can work out the relationship by plotting a graph of mass against work done

P1.4

A increases, decreases

work, force, less, easier

J, mass (kg) × gravitational field strength (N/kg) × height (m)

B change in gravitational potential

energy = mass (kg) × gravitational field strength (N/kg) × change in height (m)

$$= 30\,kg \times 10\,N/kg \times 1\,m$$
$$= 300\,J$$

C 20 J, 0.16 J, 34 J, 0.01 J

D when the student pulls the tub across the floor, she is doing work against friction; when the student pulls the tub up the ramp, she is doing work against friction and against gravity; the work done pulling the tub up the ramp will be bigger, because she has to do work against gravity and against friction

P1.5

A mass, speed

J, 0.5 × mass (kg) × (speed (m/s))²

work, elastic

B a kinetic energy (J) = 0.5 × mass (kg) ×

(speed (m/s))²
$$= 0.5 \times 0.05\,(kg) \times$$
(12.5 (m/s))²
$$= 3.9\,J$$

   b it is stationary

   c the ball is falling too fast to use a stopwatch, but using the light gates produces an accurate and precise result

C energy stored = 0.5 × spring constant (N/kg) × (extension (m))²

$$= 0.5 \times 145\,N/m \times (0.015)^2$$
$$= 0.0163\,J$$
speed² = 0.0163 J / (0.5 × 0.05 kg)
$$= 0.652\,m^2/s^2$$
speed = 0.81 m/s

P1.6

A want

dissipation

wasted, surroundings, hotter/warmer

B a dissipating energy to surroundings due to friction / air resistance

   b dissipating energy to surroundings due to friction / electric current / sound waves

   c dissipating energy to surroundings due to outside of kettle being hot / electric current / sound waves

C a if you are travelling at a steady speed, the energy in the kinetic store does not increase

   b the energy is transferred to the thermal store of the surroundings because of friction, air resistance, and heating

   c wasted energy because the energy is transferred to the surroundings and does not increase the speed

or

useful energy because if no energy was being transferred from the chemical store of the fuel the car would slow down and stop

D when you use an electric heater, energy is dissipated, or transferred to the surroundings to heat the room by an electric current; the heating element in the heater gets hot, and useful energy is transferred by heating; the heater itself gets hot, and continues to heat the room, so there is no wasted energy, because the process that you want is that the room becomes warmer

P1.7

A $\dfrac{\text{useful output energy transfer (J)}}{\text{total input energy transfer (J)}}$

100, created

friction, heating, hotter, lubricating

B a efficiency

$= \dfrac{\text{useful output energy transfer (J)}}{\text{total input energy transfer (J)}}$

$= \dfrac{400\,J}{1000\,J}$

$= 0.4$ or 40%

   b input energy transfer

$= \dfrac{\text{useful output energy transfer}}{\text{efficiency}}$

$= \dfrac{25\,J}{0.4}$

$= 62.5\,J = 63\,J$

C energy is wasted by friction, sound, and heating; so to reduce friction and sound, lubricate the moving parts; and to reduce heating by electric current, use wires with a low resistance

D efficiency = useful energy / input energy, and at the start the input energy is the gravitational potential energy; so the student is correct if her use of the ball depends on it bouncing high, because if it does not then energy has been wasted when it bounces; but because there are many ways to use a ball that do not involve bounce height, the student's statement might not be correct

P1.8

A oil

electricity, cooking/heating, lighting, microwave

less

B a kettle is more efficient because you need less input energy to produce the same amount of useful energy than with an oven; this is because you would need to heat only a smaller kettle, not a whole kettle, in addition to the water

C a **X**

b **Y** requires more energy to be supplied per minute to produce the same amount of light, so there must be more wasted energy, so it is less efficient than **X**

c **Y**

d wasted energy = input energy – useful energy

if the useful energy is the same but the input energy is larger, then more energy is wasted.

D power – a more powerful appliance will do a task in less time

efficiency – a more efficient appliance will waste less energy, so save money

P1.9

A rate

energy / work done (J)

time (s)

$\dfrac{\text{total power out (W)}}{\text{total power in (W)}} \times 100$

(W), total, (W), total, (W)

B power tells you the rate of energy transfer, and efficiency tells you the proportion of energy that is transferred usefully

a powerful device transfers energy quickly, but this might not be via the pathway that you want, so this might make it inefficient

C a useful power = $\dfrac{170\,000\,(\text{J})}{15\,(\text{s})}$

= 11 333 W (11 000 W)

b efficiency = $\dfrac{11\,000\,\text{W}}{20\,000\,\text{W}} \times 100$

= 55%

c wasted power = 20 000 W – 11 000 W

= 9000 W

P1 Practice questions

01.1 energy is transferred from a gravitational potential store [1] to a kinetic store and a thermal store [1]

01.2 energy is transferred to the thermal store of the surroundings [1]

02.1 the scientist measured the time to boil the water, and looked at the power of the kettle [1]

energy supplied = power (W) × time (s) [1]

02.2 the amount of water [1]

02.3 efficiency

= $\dfrac{\text{useful energy transfer (J)}}{\text{total input energy transfer (J)}} \times 100$ [1]

= $\dfrac{300\,000\,\text{J}}{400\,000\,\text{J}} \times 100$ [1]

= 75% [1]

02.4 insulate the kettle / make it less noisy [1]

03.1 gravitational potential energy

= mass (kg) × gravitational field strength (N/kg) × change in height (m)

= 50 kg × 10 N/kg × 1.9 m [1]

= 950 J [1]

$E_k = 0.5 \times \text{mass} \times \text{speed}^2$ [1]

speed = $\sqrt{\left(2 \times \left(\dfrac{\text{energy}}{\text{mass}}\right)\right)}$

= $\sqrt{\left(2 \times \dfrac{950\,\text{J}}{50\,\text{kg}}\right)}$

= 6.2 m/s [1]

03.2 energy has been transferred to the thermal store of the surroundings [1]

04 example answer:

calculations of work done / gravitational potential energy:

work done = force × distance

= weight × vertical height

= mass × gravity × height

Empire State Building: gravitational potential energy = 10 × 70 kg × 10 N/kg × 86 × 3 m/floor

= 1 800 000 J

The Shard: gravitational potential energy = 10 × 70 kg × 10 N/kg × 70 × 3 m/floor

= 1 500 000 J [1 for value for ESB or Shard]

so the work done in moving people to the top of the Empire State Building is larger by a factor of 1.2 [1]

power of motor = $\dfrac{\text{energy}}{\text{time}}$

Empire State Building:

power = $\dfrac{1\,800\,000\,\text{J}}{55\,\text{s}}$

= 33 000 W

The Shard: power = $\dfrac{1\,500\,000\,\text{J}}{60\,\text{s}}$

= 25 000 W [1 for value for ESB or Shard]

so the power of the motor in the Empire State Building is larger by a factor of 1.32 [1]

efficiency = $\dfrac{\text{useful energy transfer}}{\text{total input energy}}$

total input energy = $\dfrac{\text{useful energy transfer}}{\text{efficiency}}$

wasted energy = total input energy – useful energy

Empire State Building:

total input energy = $\dfrac{1\,800\,000\,\text{J}}{0.85}$

= 2 100 000 J

wasted energy = 2 100 000 J – 1 800 000 J

= 300 000 J

The Shard:

total input energy = $\dfrac{1\,500\,000\,\text{J}}{0.9}$

= 1 700 000 J

wasted energy = 1 700 000 J – 1 500 000 J

= 200 000 J [1 for wasted energy for ESB or Shard]

so The Shard motor wastes two thirds of the energy that the Empire State Building motor does [1]

P2.1

A metals, non-metals

high

lower

B a **Y, Z, X**

b the material with the lowest thermal conductivity will have the lowest rate of energy transfer through it, so its wax will melt last

c use a temperature sensor at the end of the rod to measure the time more precisely / use an extended heat source so all rods are heated evenly / use one rod at a time so the heat source heats each rod the same amount

C the thermal conductivity of air must be lower than the thermal conductivity of the loft insulation material; using material that traps air lowers the overall thermal conductivity and makes the insulation more effective

P2.2

A 1 kg, 1

longer

energy, thermometer, mass

B **X, Y, Z**

energy needed depends on mass, specific heat capacity, and the change in temperature; **Z** is double **Z** because the specific heat capacity and mass are the same but the temperature difference is double for **Z**; **X** is smaller than **Z** because the mass is doubled and the temperature difference is also doubled, so the energy is quadrupled overall, but the specific heat capacity is $\dfrac{4200}{900}$ = 4.7 times smaller, so overall it is $\dfrac{4}{4.7}$ = 0.86 times smaller than **Z**

C energy = power × time

energy needed to heat **Z** is $\dfrac{4}{(4200/900)} = \dfrac{4}{4.7}$

= 0.86, so the time taken to heat **X** would be 0.86 of the time taken to heat **Z**

D a change in thermal energy (J)

= mass (kg) × specific heat capacity (J/kg °C) × change in temperature (°C)

= 0.25 kg × 4200 J/kg °C × 11.5 °C

= 12 075 J (12 000 J)

b it would take longer / twice as long / 30 minutes

P2.3

A oil, oil, wood

loft insulation, double glazing

two, cavity wall, thick / thicker, low / lower

B oil: stove, central heating system

coal or wood: stove, fire

gas: stove, central heating system, fire

C a foam that is pumped into the gap between two walls; low thermal conductivity

b loft insulation / thick carpets / double glazing; it traps air, which has a low thermal conductivity c   aluminium foil behind radiators; it reflects radiation away from the wall and does not trap air

D   it would take $\dfrac{£2000}{£100 \text{ per year}}$ = 20 years before they start saving money with double glazing, but only $\dfrac{£175}{£25}$ = 7 years with loft insulation;

with double glazing they would save:
£100 × (30 − 20) years = £1000
with loft insulation they would save:
£25 × (30 − 7) years = £575
so double glazing would be the better choice

P2 Practice questions

01.1 **Y** [1]
01.2 **Y** [1]
01.3 lid [1] – because energy would not be transferred to the air by heating, so the insulation would be the main transfer method [1]
or
put all the cans on the same surface [1] – because energy could be transferred better through some materials than others making the test unfair [1]
02.1 to ensure that energy is transferred to the block and not to the surroundings [1]
02.2 specific heat capacity (J/kg °C)

$= \dfrac{45 \text{ J}}{1 \text{ kg} \times 20\,°C}$ [1]

= 2.3 (2.25) J/kg °C [1]
02.3 the measured value of the specific heat capacity is too high [1]
because the student needs to transfer more energy because it is heating the surroundings [1]
03.1 when the temperature difference doubles from 33 °C to 66 °C, [1] the thickness of insulation doubles from 38 mm to 76 mm [1]

$\left(\text{or } \dfrac{66}{33} = \dfrac{76}{38} \right)$

03.2 the materials used in modern sleeping bags have a lower thermal conductivity, [1] so the thickness needed can be less and still result in the same rate of transfer of energy [1]

P3.1

A coal, oil, gas
will
will not
biofuels, methane, ethanol
nuclear, more
B a biofuels / coal
    b wood
C a a fuel produced from recently living plants, or animal waste
    b it is renewable because you can produce more of it in a reasonable timescale
D a uranium atoms are unstable and can be split in two

b i 1 tonne of animal manure releases 12 MJ × 1000 = 12 000 MJ
mass of fissile uranium needed
$= \dfrac{12\,000 \text{ MJ}}{77\,000\,000 \text{ MJ/kg}} = 1.6 \times 10^{-4} \text{ kg}$

    ii $\dfrac{1}{0.007}$ = 143 times as much uranium
metal needed as fissile uranium
143 × 1.6 × 10⁻⁴ kg = 0.023 kg

P3.2

A turbine
wave
turbines, hydroelectric
tidal
environment / habitats
B a both rely on moving water to drive a turbine / generator to generate electricity
    b both rely on moving air to drive a turbine / generator to generate electricity
C water behind a dam high up falls through turbines and generators to generate electricity
D a once the power station is built it costs very little to run
    b a tidal barrage is very expensive to build compared with wave generators

P3.3

A generate, electricity, small, cheap, nothing
heat
mirrors, electricity
radioactive, water, steam, turbines
B all three use the light/radiation from the Sun; a solar heating panel uses radiation from the Sun to heat water, but a solar power tower and a solar cell panel generate electricity
C advantages: you can generate electricity in remote places; you do not need to be connected to a power station
disadvantages: solar cells convert less than 10% of solar radiation into electricity; solar panels work only on sunny days; solar cells are very expensive to buy; you need lots of solar cell panels to generate enough electricity to be useful

P3.4

A greenhouse, warming, acid rain
more, radioactive, expensive
do not, remote, animals, plants, expensive
B use carbon capture and storage to reduce effect of greenhouse gas carbon dioxide; remove sulfur to reduce the effect of acid rain
C wind
tidal
wind, solar
hydro
solar
wind, tidal, hydro, solar
D a a nuclear power station does not produce greenhouse gases / the energy produced from 1 kg of nuclear fuel is

much greater than the energy from 1 kg of a fossil fuel
    b there is less danger of an accident from a fossil fuel power station / radioactive waste lasts thousands of years

P3.5

A coal, pumped, expensive, decommission carbon capture, cheap, expensive, energy
B a the constant minimum amount of electricity that is needed
    b by pumping water up into a reservoir at a hydroelectric power station
    c solar power cannot provide the base load / cannot provide electricity at times of peak demand
C a capital costs are costs of building and decommissioning the station; running costs are daily costs of running the station
    b will decrease as technology improves / use increases
D wind farm turbines are at the top of a pole so difficult and expensive to repair, so running costs are greater

P3 Practice questions

01.1 coal, oil, gas [1 for all three]
01.2 wind: uses air to turn a turbine, a turbine turns a generator [1]
waves: uses water to turn a turbine, a turbine turns a generator [1]
tides: uses water to turn a turbine, a turbine turns a generator, reliable resource [1]
02 the uranium in the fuel rods releases energy; [1] the core gets hot; [1] the coolant pumped through the core gets hot; [1] the hot coolant turns the water to steam; [1] the steam drives a turbine; [1] the turbine drives a generator, which produces electricity [1]
03.1 total percentage = percentage for oil + coal + gas
= 1% + 31% + 46% [1]
= 78% [1]
03.2 e.g., burning fossil fuels causes acid rain / produces greenhouse gases; [1] burning biofuels produces greenhouse gases but not acid rain; [1] greenhouse gases contribute to global warming; [1] sulfur can produce acid rain / sulfur dioxide [1]
03.3 one from: wind, waves, tidal, geothermal, hydroelectric, solar [1]
03.4 appropriate advantage, e.g., does not produce greenhouse gases, cheap to run [1]
appropriate disadvantage, e.g., expensive to install, unreliable [1]
04 **Sasha's reply:** I don't think we should, because they produce radioactive waste [1] and they are expensive to decommission. [1]
Dev's reply: I don't think we should, because they are expensive to install [1] and they can harm wildlife. [1]

05.1 a solar heating panel [1] or a solar cell panel [1]

05.2 total area of roofs in the UK = 25 000 000 × 140 m² = 3 500 000 000 m² [1]

output = 250 W/m² × 3 500 000 000 m² = 875 000 000 000 W [1]

percentage needed =

$$\frac{5\ 500\ 000\ 000\ W}{875\ 000\ 000\ 000\ W} \times 100 \text{ [1]} = 0.6\% \text{ [1]}$$

05.3 suitable comment with reasoning, e.g., only a small proportion of houses needed, [1] but the calculation is based on the maximum output; [1] the cells would produce much less electricity than the maximum most of the time, [1] so the proportion of houses needed would be much bigger [1]

P4.1

A symbol, cells
rate
(C), $\dfrac{\text{current (A)}}{\text{time (s)}}$

B

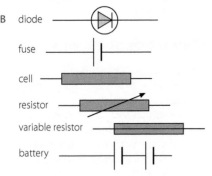

diode

fuse

cell

resistor

variable resistor

battery

C a Charge is measured in coulombs/C. or Current is measured in amperes.
b Time is always measured in seconds.
c The current before a component is the same as the current after a component.
d Current is the rate of flow of charge/ charge flowing per second.

D 2 minutes = 120 seconds

current = $\dfrac{15\,C}{120\,s}$ = 0.13 A (0.125 A)

P4.2

A energy, energy, volts
(V), $\dfrac{\text{energy (J)}}{\text{charge (C)}}$
(Ω), $\dfrac{\text{potential difference (V)}}{\text{current (A)}}$
directly proportional, reverse

B a drawing of series circuit containing battery, lamp, and ammeter; voltmeter in parallel with lamp
b the ammeter is in series with the lamp because it measures the current through it; the voltmeter is in parallel with the

lamp because it measures the difference in energy either side of it
c they would show a negative value of current

C charge = $\dfrac{\text{energy}}{\text{p.d.}}$

$= \dfrac{200\ J}{12\ V} = 17\,C\ (16.7\,C)$

D a $I = \dfrac{V}{R}$

$= \dfrac{12}{30} = 0.4\ A$

$Q = I\,t$, so $t = \dfrac{Q}{I}$

$= \dfrac{600}{0.4} = 1500\ s$

b its temperature does not change / it obeys Ohm's law / it is an ohmic conductor

P4.3

A increases
low, high, forward
decreases, increases

B a component **Z**: the current is proportional to the p.d.
b component **X**: the ratio of p.d. to current is higher at higher p.d.s / gradient of graph decreases at higher p.d.
c a diode: it has a low resistance for positive p.d. after about 0.7 V but a very high resistance for negative p.d.

C light-dependent resistor: dark, light
thermistor: cold, hot

D as the electrons move through the filament wire they collide with the ions/atoms, which vibrate more as the wire gets hotter; the increasing collisions make it harder for the electrons to move through the wire

P4.4

A current, potential difference, adding
add
increases, less

B a **X** – 0.2; **Y** – 0.2; **Z** – 0.2; cell – 3; bulb 1 – 1.5; bulb 2 – 1.5
b the current everywhere in a series circuit is the same; the potential differences across the bulbs add up to the potential difference across the cell; the bulbs are identical so the potential difference across each of them is the same

C resistance of bulb 1 or bulb 2 = $\dfrac{1.5\ V}{0.2\ A}$ = 7.5 Ω

total resistance of circuit = $\dfrac{3\ V}{0.2\ A}$ = 15 Ω,

or 2 × 7.5 Ω = 15 Ω

D the bulbs are not ohmic conductors / the resistance of the bulb is not constant / increases with temperature

P4.5

A potential difference, current, currents
smaller
(A) $\dfrac{\text{p.d. (V)}}{\text{resistance (Ω)}}$
decreases, increases

B a **X** – 0.2; **Y** – 0.2; **Z** – 0.4; cell – 9; bulb 1 – 9; bulb 2 – 9
b the current at **Y** will be the same as the current at **X** because the bulbs are identical and the potential difference across them is the same; the current at **Z** is the sum of the currents at **X** and **Y**; the potential difference across each bulb is the same as that across the cell

C resistance of bulb 1 or bulb 2 = $\dfrac{9\ V}{0.2\ A}$ = 45 Ω

total resistance of circuit = $\dfrac{9\ V}{0.4\ A}$ = 22.5 Ω

D increase the p.d. of the cell / add another cell

P4 Practice questions

01.1 ammeter, [1] voltmeter [1]
01.2 to change the p.d. across the bulb / to change the current through the bulb / to obtain a range of values of p.d. and current [1]
02.1 close one switch, if all the bulbs go off it is in series, if only one goes off it is in parallel [1]
02.2 10 Ω + 15 Ω = 25 Ω [1]
02.3 current = $\dfrac{\text{p.d.}}{\text{resistance}}$

$= \dfrac{12\ V}{15\ Ω}$ and $\dfrac{12\ V}{10\ Ω}$

$= 0.8\ A$ [1] and 1.2 A [1]
02.4 total current = 0.8 A + 1.2 A
$= 2\ A$ [1]

resistance = $\dfrac{\text{p.d.}}{\text{current}}$ [1]

$= \dfrac{12\ V}{2\ A} = 6\ Ω$ [1]

03.1 current (A) = $\dfrac{\text{charge (C)}}{\text{time (s)}}$ [1]

$= \dfrac{20\ C}{40\ s} = 0.5\ A$ [1]

03.2 the current flowing through a conductor is proportional to the p.d. across it [1] as long as the temperature stays constant [1]
03.3 resistance (Ω) = $\dfrac{\text{p.d. (V)}}{\text{current (A)}}$ [1]

$= \dfrac{9\ V}{0.5\ A} = 18\ Ω$ [1]

03.4 the direction of motion of the fan would reverse [1]
03.5 doubling the p.d. doubles the energy per charge (W = QV) [1]

and also doubles the current $\left(I = \dfrac{V}{R}\right)$ [1]

so twice as much charge flows per second and each charge transfers twice as much energy, so the energy per second is multiplied by 4 [1]

04.1 as the light level changes, the resistance of the LDR changes, [1] so the p.d across the LDR changes; [1] the voltmeter measures the p.d. across R_1, which is the p.d. of the cell minus the p.d. across the LDR, so that also changes [1]

04.2 example answer:
in the light:
total resistance = $100\,\Omega + 500\,\Omega = 600\,\Omega$ [1]
$$\text{current} = \frac{\text{p.d.}}{\text{resistance}}$$
$$= \frac{6\,V}{600\,\Omega} = 0.01\,A\ [1]$$
p.d. across R_2 = current × resistance
$$= 0.01\,A \times 500\,\Omega$$
$$= 5\,V\ [1]$$
in the dark:
total resistance = $100\,\Omega + 500\,000\,\Omega$
$$= 500\,100\,\Omega$$
$$\text{current} = \frac{\text{p.d.}}{\text{resistance}}$$
$$= \frac{6\,V}{500\,100\,\Omega}$$
$$= 0.000012\,A\ [1]$$
p.d. across R_2 = current × resistance
$$= 0.000012\,A \times 500\,\Omega$$
$$= 0.006\,V\ [1]$$
in the light, most of the p.d. is across the resistor; in the dark, most of the p.d. is across the LDR [1]

P5.1

A one, reverses
−230V, +230V, zero/0
network
maximum, zero, period (for one cycle)
Hz, $\dfrac{1}{\text{period (s)}}$

B d.c. – straight horizontal line; a.c. – goes up and down about 0V

C **a** 325V
b find the time for one cycle, then use the equation: frequency = $\dfrac{1}{\text{period}}$
c period = 0.02 s, f = $\dfrac{1}{\text{period}} = \dfrac{1}{0.02\,s} = 50\,Hz$
d live, the p.d. on the graph varies between positive and negative values, but the neutral wire is 0V
e the value of 230V is the effective voltage because the p.d. varies / it is only at the peak voltage at 2 points in the cycle

P5.2

A plastic, insulator
copper, insulating
brown, blue, green, yellow
longest / centre, casing

B **a** live, **brown**, carries the current to make an appliance work, **230**
neutral, blue, **completes the circuit to make the appliance work, 0**

earth, **green/yellow, safety / connected to earth to prevent you getting a shock**, 0
b similarity: both plastic / insulators
difference: wire insulation is soft / flexible, but plug casing is stiff / rigid

C **a** if you touch a case that is live / connected to the live wire then a current will flow through you
b if there is a fault the fuse melts and breaks the circuit; the current flows through the earth wire and not through you
c any appliance with a plastic case, e.g. hairdryer
d nothing / it would not work; there is no potential difference between the earth and neutral wires

P5.3

A energy
(J), power (W) × time (s)
(W), potential difference (V) × current (A)
(A), $\dfrac{\text{power (W)}}{\text{p.d. (V)}}$

B **a** power is the rate of transfer of energy / energy transferred per second
b the bigger the power rating, the bigger the fuse needed; electrical power = V × I, so if p.d. is the same, bigger power means bigger current

C 15 kW = 15 000 W
2.5 hours = 9000 s
energy transferred = 15 000 W × 9000 s
= 140 000 000 J (135 000 000 J)

D **a** current = $\dfrac{1000\,W}{230\,V}$
$$= 4.35\,A\ (= 4.4\,A)$$
b 5 A, a smaller value fuse would melt, and a higher value fuse would not protect the appliance

E resistance = $\dfrac{P}{I^2}$
$$= \dfrac{350\,W}{(1.5\,A)^2}$$
$$= 160\,\Omega\ (156\,\Omega)$$

P5.4

A (C), current (A) × time (s)
(J), p.d. (V) × charge (C)
heat up
energy, energy

B the electrons collide with the vibrating atoms / ions in the wire and transfer energy to them, so they vibrate more, so the wire heats up

C **a** 4 mA = 0.004 A
charge = 0.004 A × 0.1 s = 0.0004 C
b energy transferred = 300 000 C × 230 V
= 69 000 000 J

D a very small amount of energy is **dissipated** in the wires in the circuit, so the energy transferred to the components is slightly less than the energy transferred by the battery

P5.5

A energy
(J), power (W) × time (s)
(J), efficiency × energy supplied (J)
(W), efficiency × power supplied (W)

B **a** power = 230V × 15 A
$$= 3450W = 3500W\text{ to 2 significant figures}$$
b time

C **a** 30 minutes = 1800 s
energy transferred = 2000 W × 1800 s
$$= 3\,600\,000\,J$$
b 70% = 0.7 as a decimal fraction
4000 kJ = 4 000 000 J
useful energy = 0.7 × 4 000 000 J
$$= 2\,800\,000\,J$$
c efficiency = $\dfrac{\text{useful energy}}{\text{total energy transferred}} \times 100$
$$= \dfrac{2\,800\,000\,J}{3\,600\,000\,J} \times 100$$
$$= 78\%$$

P5 Practice questions

01.1 **A**: green/yellow; **B**: brown; **C**: blue [all correct – 2, one correct – 1]
01.2 **A** [1]
01.3 **B** and **C** [1]
01.4 National Grid [1]
02 hard plastic, casing, rigid insulator
flexible plastic, **insulation for wire, flexible insulator**
copper, **wire, conductor**
03 current = $\dfrac{\text{power}}{\text{p.d.}}$ [1]
$$= \dfrac{1100\,W}{230\,V}\ [1]$$
$$= 4.78\,A\ (= 4.8\,A)\ [1]$$
so she needs a 5 A fuse [1]
04.1 2 min = 120 s
energy transferred = power × time [1]
$$= 800\,W \times 120\,s\ [1]$$
$$= 96\,000\,J\ [1]$$
04.2 time = 6 min = 360 s
charge = current × time [1]
$$= 5.2\,A \times 360\,s\ [1]$$
$$= 1872\,C = 1900\,C\ (\text{to 2 s.f.})\ [1]$$
04.3 time = 24 × 60 × 60 = 86 400 s [1]
energy transferred = power × time
$$= 420\,W \times 86\,400\,s\ [1]$$
$$= 36\,288\,000\,J$$
$$= 36\,000\,000\,J\ (\text{to 2 s.f.})\ [1]$$
05.1 time = 4 × 2 ms = 8 ms [1]
8 ms = 0.008 s [1]
frequency = $\dfrac{1}{\text{period}}$ [1]
$$= \dfrac{1}{0.008\,s}$$
$$= 125\,Hz\ [1]$$
05.2 the peak p.d. is 15V, but mains is 325V / much higher [1]
the frequency is 125 Hz, but mains frequency is 50 Hz [1]

06 3 MJ = 3 000 000 J = useful energy
transferred [1]
in 1990, energy per cycle = 2.7 kWh
2.7 kWh = 2.7 kW × 1000 (W/kW) × 3600 (s/h)
      = 9 720 000 J [1]

$$\text{efficiency} = \frac{\text{useful energy}}{\text{total energy transferred}} \times 100$$

$$= \frac{3\,000\,000\,\text{J}}{9\,720\,000\,\text{J}} \times 100$$

= **30.8%** [1]
in 2010, energy per cycle = 1.4 kWh
1.4 kWh = 1.4 kW × 1000 (W/kW) × 3600 (s/h)
      = 5 040 000 J [1]

$$\text{efficiency} = \frac{3\,000\,000\,\text{J}}{5\,040\,000\,\text{J}} \times 100$$

= **59.5%** [1]
the efficiency of dishwashers has nearly
doubled in 20 years [1]

P6.1

A mass, volume, kg/m³

$$\text{kg/m}^3, \frac{\text{mass (kg)}}{\text{volume (m}^3)}$$

density, volume, mass, density
mass, volume
less than

B $\text{density} = \dfrac{55\,\text{kg}}{0.07\,\text{m}^3} = 786\,\text{kg/m}^3$

C an object floats if its density is less than the
density of water; the ship has a much larger
mass, but also a larger volume than the
pebble; the ship's density is less than that of
water (the large amount of air inside the ship
reduces the total density of the ship), but
the density of the pebble is greater than the
density of water, so the pebble sinks

D measure the mass of each object with a
digital balance; measure the volume of the
stone cube using a ruler; put the modelling
clay into a known volume of water in a
measuring cylinder; observe the volume
increase; this gives the volume of the
modelling clay

E mass of **X** = density of **X** × volume of **X**
      = 1.5 g/cm³ × 10 cm³
      = 15 g

volume of **Y** with this mass = $\dfrac{\text{mass}}{\text{density}}$

$$= \frac{15\,\text{g}}{5\,\text{g/cm}^3} = 3\,\text{cm}^3$$

P6.2

A gas, liquid, solid
solid, gas,
mass

B in both arrangements, the particles are
touching; density is the mass per unit volume;
for a given mass, the volumes will be similar,
and so will the densities

C a

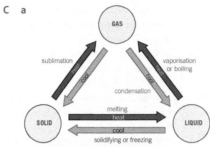

b the particles have more kinetic energy,
because they are moving faster, and
more potential energy because they are
slightly further apart

D a it decreases

b the mass of water in the air + the mass of
water in the bowl = the original mass of
water in the bowl

c no new substances are formed

P6.3

A melts, freezes, boils, condenses,
horizontal
boiling, vaporisation

B This process happens at the boiling point of
the liquid – *boiling*
The mass does not change – *boiling,
evaporation*
The particles escape only from the surface of
the liquid – *boiling, evaporation*
This process happens at or below the boiling
point of the liquid – *evaporation*

C a dotted line: melting point
first curved section: solid
horizontal section: solid + liquid
second curved section: liquid

b a mirror image of the graph in part **a**,
with the line starting on the *y*-axis
where it ends on the graph in part a; the
horizontal line should be in the same
place on the temperature axis

c label 'freezing point', instead of 'melting
point', on the temperature axis

P6.4

A increases
attraction
kinetic, potential
colliding

B the kinetic energy of particles in a gas; the
energy of vibration of the particles in a solid;
the gravitational potential energy of the
particles in a liquid

C particles repeatedly collide with the surface of
the container

D a the kinetic energy of the particles in the
liquid increases, but the kinetic energy of
the particles in the solid decreases; there
is no change to the potential energy of
the particles in the solid or liquid

b if a substance is melting or boiling /
vaporising, the energy will break the
bonds between the particles but not
increase the temperature; the internal
energy increases because the potential

energy increases, but the kinetic energy,
and so the temperature, does not increase

P6.5

A energy, temperature
1 kg, temperature
(J), mass (kg) × specific latent heat (J/kg)
melt, boil

B latent heat is the energy needed to change
the state of a substance; specific latent heat is
the energy to change 1 kg of the substance;
latent heat is measured in J; specific latent
heat is measured in J/kg

C 334 kJ/kg = 334 000 J/kg
thermal energy for melting ice =
      334 000 J/kg × 0.03 kg = 10 020 J

D when you heat a substance and it changes
state, you increase the potential energy but
not its kinetic energy; this requires more
energy per kilogram for a liquid to a gas
than for a solid to a liquid because you are
completely breaking the bonds between
particles

P6.6

A pressure
increases, more, increases
random

B Brownian motion / observing smoke
particles that you can see being moved by air
molecules that you cannot see

C if the temperature is higher, the molecules are
moving faster, so they collide more often with
the walls of the container; so the gas exerts a
bigger force on the walls of the container as
the gas is heated

D a the mass of gas / volume of the
container; if either changed, the pressure
change would be due to more / fewer
particles, or fewer / more collisions
between the gas particles and the
container

b it would flatten off, or curve downwards,
because the number of gas particles
would decrease due to the leak, so the
pressure would be less

P6 Practice questions

01 energy to change the state of a material. [1]

02.1 freezing is the process of particles moving
into a fixed pattern, with bonds forming
between them; [1] melting is the process
of particles moving from a fixed pattern
into an irregular pattern, with bonds
breaking / becoming weaker between
them [1]

02.2 vaporisation is the process of particles
changing from a liquid to a gas (bonds
breaking totally) at temperatures below
the boiling point; [1] boiling only occurs at
the boiling point [1]

03.1 $\text{density} = \dfrac{\text{mass}}{\text{volume}}$ [1]

$= \dfrac{2000 \text{ kg}}{0.5 \text{ m}^3}$ [1]

$= 4000 \text{ kg/m}^3$ [1]

03.2 they would all sink [1] because they all have densities greater than water [1]

04.1 70 °C [1]

04.2 9 mins [1]

04.3 0 mins [1]

04.4 it is increasing [1] because the potential energy is increasing as bonds are broken, but the kinetic energy does not change because the temperature does not change [1]

05.1 D before B, [1] B before C, [1] C before A, [1] A before E [1]

05.2 if the student does not collect all the water from the melting of the ice, the mass measured will be less than it should be [1]

$\text{specific latent heat} = \dfrac{\text{energy}}{\text{mass}}$ [1]

so if the mass is smaller, then the specific latent heat will be bigger [1]

06 mass = 2 g = 0.002 kg

thermal energy for a change of state

= 0.002 kg × 334 000 J/kg

= 668 J [1]

temperature change = 20 °C [1]

change in thermal energy

= 0.002 kg × 4200 J/kg °C × 20 °C

= 168 J [1]

total energy = 668 J + 168 J

= 836 J = 840 J [1]

P7.1

A unstable, stable

alpha, beta, gamma

random

B example: throwing a die, or tossing a coin; you cannot predict the number / whether it will be heads or tails

C alpha, **α, nucleus**

gamma, γ, nucleus

beta, β, nucleus

D **a** the rock is emitting waves or particles, which are detected by the Geiger counter

b put a piece of paper between the sample and the Geiger counter, and if the count rate goes down the source is emitting alpha radiation

c the atoms in a sample that produces a reading are unstable / emitting radiation, but those in a sample that does not produce a reading are not

P7.2

A alpha, large

plum pudding

massive / small, positively

B **a** 2 lines with angle θ > 90°

b most alpha particles went through, so most of the atom was empty space; a few came back, so there is a very small, dense, positively charged nucleus at the

centre of the atom, which repels the positive alpha particles

C in both models there is a small, dense, positively charged nucleus in the centre of the atom; in Rutherford's model, the electrons orbit the nucleus, but in Bohr's model, the electrons can only orbit at certain distances, which are called energy levels / shells

P7.3

A same, different, same, different

two / 2, two / 2, four / 4, two / 2

neutron, proton, electron, does not change, one / 1

B **a** $^{14}_{6}\text{X}$ $^{12}_{6}\text{X}$

b isotopes have the same number of protons and different numbers of neutrons; the bottom number is the number of protons, so the two isotopes are the ones with the same bottom number

C in alpha decay, a nucleus loses two protons and two neutrons; in beta decay, the number of nucleons stays the same, but the number of protons increases by one, and the number of neutrons decreases by one

D $^{226}_{88}\text{Ra} \rightarrow {}^{222}_{86}\text{Ra} + {}^{4}_{2}\text{He}$

$^{218}_{84}\text{Po} \rightarrow {}^{218}_{85}\text{At} + {}^{0}_{-1}\text{e}$

E in both cases, an uncharged particle / wave is emitted, and the type of element does not change; in gamma emission, an electromagnetic wave is emitted, and the mass does not change; in neutron emission, a particle is emitted and the mass decreases

P7.4

A paper, aluminium, lead

a few cm, about 1 m, infinite

2 / two, 2 / two, fast, electron, electromagnetic

most, least

ionise, damage

B set up the Geiger counter so that is detecting radiation from the sample; put a sheet of paper between the counter and the sample: if the reading goes down, the sample is emitting alpha; put a sheet of aluminium between the counter and the sample: if the reading goes down, the sample is emitting beta; if there is no change with aluminium, the sample is emitting gamma

C **a**

It has an infinite range in air	γ	It consists of a fast moving electron	β
It is moderately ionising	β	It is the most ionising	α
It has a range of about 1 m in air	β	It has a range of a few cm in air	α
It consists of two protons and two neutrons	α	It consists of electromagnetic radiation	γ

b it is the least ionising – γ

D irradiation means that the source of radiation was outside the strawberries; the strawberries do not become radioactive; eating them will not put radioactive material in the body, so will not cause cancer

P7.5

A halve, decreases, halve, 2^n

B count rate is the reading measured with a Geiger counter; activity is the number of decays per second

C Half-life is the time for the number of unstable nuclei to halve.

Half-life is the time for the activity to halve.

D **a** 6 days

b initial activity = 80 counts/s

after 3 half-lives $= \dfrac{80}{2^3} = 10$ counts/s

c 5 counts/s after 4 half-lives, which is 24 days

suitable method, e.g.:

counts: 80 – 40 – 20 – 10 – 5 means you need to halve 80 four times

d the ratio of net decline $= \dfrac{1}{2^6} = \dfrac{1}{64}$

P7 Practice questions

01.1 an alpha particle is a helium nucleus / two protons and two neutrons; a beta particle is a fast-moving electron; a gamma ray is an electromagnetic wave [1]

alpha is absorbed by paper; beta by aluminium; gamma by lead and concrete [1]

alpha travels a few cm; beta about 1 m; gamma has an infinite range [1]

01.2 when radiation ionises, it loses energy; [1] alpha is very ionising, so does not travel far because it loses a lot of energy, [1] but gamma hardly ionises so has a very large range [1]

02 example answer: J J Thompson thought the model was **A**; [1] Geiger and Marsden did an experiment firing alpha particles at gold foil [1] and found that most of the alpha particles went through the foil [1] but some were scattered back / through large angles [1]; this led Rutherford to develop the model in **C**; [1] we now use the model in **B** which explains why electrons do not spiral in to the nucleus [1]

03.1 alpha [1]

03.2 gamma [1]

03.3 beta [1]

03.4 beta or gamma [1]

04.1 the number of unstable nuclei [1]

04.2 initial activity = 2000 counts/min [1]

activity after 5 half-lives

$= \dfrac{2000}{2^5}$ [1]

= 62.5 counts/min [1]

= 63 counts/min (cannot have half a count) [1]

05.1 ionising radiation can damage or kill cells [1]

05.2 the nucleus of technetium-101 has two more neutrons than the nucleus of technetium-99 [1]

05.3 $^{101}_{43}\text{Tc} \rightarrow ^{101}_{44}\text{Ru} + ^{0}_{-1}\beta$ [2]

05.4 the half-life of technetium-101 is too short / it would not emit enough radiation to detect / may not have reached the kidneys before it has decayed too much; [1] it emits beta radiation, which is ionising and could cause cancer; [1] beta radiation would be stopped by the body tissue and would not be detected by the camera outside the body [1]

P8.1

A distance
magnitude, direction, magnitude
direction, magnitude

B a e.g. temperature / distance; it has no direction
b because an arrow can show a direction

C a 20 cm east
b 90 cm

D suitable scale diagram with scale shown (e.g. 4 cm arrow upwards and 1 cm arrow to the left, scale 5 N/cm)

P8.2

A size, shape, stationary, newtons
contact
equal, opposite

B a e.g. a football changes shape when it is headed
b e.g. a car accelerates when it is on the motorway
c e.g. a cyclist starts moving when she pedals

C contact forces are exerted when objects are in contact, e.g. friction; non-contact forces are exerted whether or not the objects are in contact, e.g. gravity

D a when two objects interact with each other, they exert equal and opposite forces on each other
b the forces acting are the force of the Earth on you, and the force of the chair on you, so both forces are acting on you, not on different objects.
c when you walk, you exert a force on the floor, and the floor exerts a force on you, so you move forward
d when the fuel burns, exhaust gases leave the bottom of the rocket; the pair of forces are the force of the gas on the rocket and the force of the rocket on the gas; when the force of the gas on the rocket upwards exceeds the force of the Earth on the rocket (weight), the rocket takes off

P8.3

A same
zero
bigger, zero
add, difference
free-body

B if the cyclists is moving at a constant speed then friction/air resistance balances the force from the cyclist, so the resultant force is zero

C a stationary, moving at a steady speed
b 7, 3, **4, to the left**
10, 20, **10, to the right**
80, 150, **70, to the right**

D a arrow pointing up; arrow pointing down; both arrows shown on the dot
b you are in equilibrium/at rest; the arrows are equal length but in opposite directions

P8.4

A point
centre
below
line

B the racing car is not as high so that its centre of mass is lower, so it is less likely to topple over when it goes round a corner at high speed

C circle with dot in the centre, rectangle with dot in the centre, triangle with dot in the centre

D a **a** – vertical arrow pointing down to the right of the pivot; **b** – vertical arrow pointing down on the pivot; **c** – vertical arrow pointing down to the left of the pivot; all arrows start at centre of mass
b to topple the lorry, the weight has to act to the left of the pivot; the centre of mass must be to the left of the pivot; so a wind will need to move the centre of mass by a big distance, so the wind needs to be very strong

P8.5

A scale
resultant
protractor
origin

B you use a parallelogram of forces when the forces are acting at an angle to each other, but not when they act in a single line

C a diagram with one arrow two-thirds the length of the other, and the angle between them = 30°
b 9.7 (10)

D a resultant drawn using the parallelogram of forces

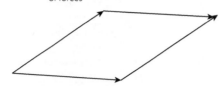

b i length of resultant increases; angle between resultant and tug A increases
ii the resultant would be the force of tug B

P8.6 Resolution of forces

A 90°
rectangle, diagonal
equilibrium
rest

B a 90°
b the magnitude of the normal force when you are sitting on the flat ground is bigger than the magnitude of the normal force when you are sitting on the hillside

C rectangle drawn; correct horizontal and vertical forces (8.6 N and 5 N)

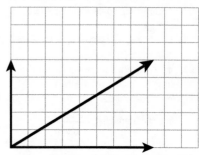

D a components drawn along the slope and perpendicular to the slope

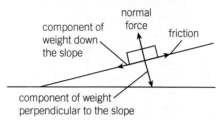

b angle of the ramp and weight of the block; then the student can draw a scale diagram to find the force down the slope, which is equal to the frictional force
c the height will be smaller; the frictional force is less, so the component of the weight down the slope will need to be much smaller

P8 Practice questions

01 force has direction but temperature does not [1]
02 friction, tension, air resistance [all correct – 2, 2 or 1 correct – 1]
03.1 → 3N; [1] 0N; [1] ← 1N [1]
03.2 box B; [1] there is zero resultant force [1]
03.3 accelerate to the right; [1] move to the left at 3 m/s; [1] move to the left and slow down [1]
04.1 walking in a straight line [1]
04.2 walking in a circle / return to the start [1]
05 they are the same [1]
06.1 it is below the suspension point [1]
06.2 dot with arrow downwards labelled 'weight' [1] and arrow up and to the right at 30° to vertical labelled 'tension'; [1] vertical component of tension drawn upwards; [1] scale used to produce answer: T = 57 N [1]

07.1 diagram to scale with two arrows at 10° to the horizontal pointing left and right [1] arrow downwards labelled 'weight'; [1] parallelogram drawn; [1] scale used to find weight = 700 N [1]

07.2 resultant of the two forces in the parallelogram balances the weight; [1] if the angle is very small the tension in the wire will be much bigger, [1] so the wire may break [1]

P9.1

A m/s, $\dfrac{\text{distance (m)}}{\text{time (s)}}$

stationary, steady / constant speed

gradient

B a speed $= \dfrac{11\,\text{m}}{2.7\,\text{s}} = 4.1\,\text{m/s}$

b steady / constant speed

C a distance travelled = 5.5 m/s × 15 s

= 83 m

b time $= \dfrac{70\,\text{m}}{5.5\,\text{m/s}} = 13\,\text{s}$

D in section **A**, the object is travelling at a steady speed; in section **B**, the object is stationary; the line is straight with a gradient; the line has zero gradient

in **A**, speed $= \dfrac{\text{distance}}{\text{time}}$

$= \dfrac{20\,\text{m}}{4\,\text{s}}$

= 5 m/s, constant speed

in **B**, speed = 0 m/s, constant speed

P9.2

A direction

vector, scalar

(m/s^2), $\dfrac{\text{change in velocity (m/s)}}{\text{time (s)}}$

slows down

velocity

B a speed has magnitude only, velocity has magnitude and direction

b 10 m/s north, 38 mph south

C a speed $= \dfrac{\text{distance}}{\text{time}}$

$= 2 \times \pi \times \dfrac{r}{t}$

$= 2 \times 3.14 \times \dfrac{25\,\text{m}}{20\,\text{s}}$

= 7.9 m/s

b the velocity has a constant magnitude but the direction is constantly changing, so it is accelerating towards the centre of the circle

c the speed is the same, but the velocities are in opposite directions

D a change in velocity = 100 m/s – 0 m/s

= 100 m/s

acceleration $= \dfrac{100\,\text{m/s}}{30\,\text{s}}$

$= 3.3\,\text{m/s}^2$

b change in velocity = 140 m/s – 155 m/s

= –15 m/s

acceleration $= \dfrac{-15\,\text{m/s}}{120\,\text{s}}$

$= -0.125\,\text{m/s}^2$

P9.3

A motion sensor, gradient, zero

positive gradient

negative gradient

distance travelled

B advantage: a motion sensor can plot the graph as the object is moving

disadvantage: it needs to be connected to a computer at the time of analysis

C accelerating; decelerating; moving at a steady speed

D a acceleration $= \dfrac{(18\,\text{m/s} - 0\,\text{m/s})}{15\,\text{s}}$

$= 1.2\,\text{m/s}^2$

b total distance travelled

$= \dfrac{1}{2}(18\,\text{m/s} \times 15\,\text{s}) + (18\,\text{m/s} \times 30\,\text{s}) +$

$\dfrac{1}{2}(18\,\text{m/s} \times 15\,\text{s})$

= 810 m

c the change in velocity is the same; the time to change is the same; so the acceleration is the same magnitude but negative (–1.2 m/s²)

P9.4

A gradient

gradient

area under

gradient, tangent

B a horizontal line on a distance-time graph means that the object is stationary; a horizontal line on a speed-time graph means that the object is moving at a constant speed, but if the speed is zero the object is stationary

C zero, zero, positive

D a gradient drawn on graph; speed = 10 m/s

b gradient drawn on graph; speed = 8.5 m/s

c graph **B**; the speed is increasing at a steady rate

P9 Practice questions

01 7.4; [1] 0.24; [1] 26 [1]

02 acceleration $= \dfrac{\text{change in speed}}{\text{time}}$ [1]

$= \dfrac{0.5\,\text{m/s}}{0.05\,\text{s}}$ [1]

$= 10\,\text{m/s}^2$ [1]

03.1 1000 m or 3000 m [1]

03.2 speed $= \dfrac{\text{distance}}{\text{time}}$

$= \dfrac{2000\,\text{m}}{600\,\text{s}}$ [1]

= 3.3 m/s [1]

03.3 using a stopwatch [1] and a trundle wheel [1]

04.1 **A** to **B**: acceleration $= \dfrac{\text{change in speed}}{\text{time}}$

$= \dfrac{7\,\text{m/s}}{10\,\text{s}}$ [1]

$= 0.7\,\text{m/s}^2$ [1]

between **B** and **C**: zero; [1] between **C** and **D**: acceleration is half / 0.35 m/s², [1] and negative / it is deceleration [1]

04.2 the cyclist's speed is changing so you cannot use distance = speed × time [1]

04.3 distance $= \dfrac{1}{2}(7\,\text{m/s} \times 10\,\text{s}) + (7\,\text{m/s} \times 20\,\text{s})$

$+ \dfrac{1}{2}(7\,\text{m/s} \times 20\,\text{s})$ [1]

= 245 m [1]

04.4 the time it takes to travel half the circle: [1] the acceleration is the change in velocity divided by the time; [1] the velocity changes by 10 m/s, so you need time to calculate acceleration [1]

05 tangent drawn; [1] values of displacement and time used from tangent [1]

speed $= \dfrac{(60\,\text{m} - 0\,\text{m})}{(0.88\,\text{s} - 0.24\,\text{s})}$

= 9.4 m/s [1]

P10.1

A increase, mass

smaller

N, mass (kg) × acceleration (m/s²)

rest, constant / uniform

B if the force acting on an object increases, so does the acceleration; if the mass of an object increases but the force stays constant, the acceleration will decrease

C inertial mass is a measure of the resistance an object has when a force is exerted on it

D a mass = 105 g = 0.105 kg

resultant force = 0.105 kg × 3.20 m/s²

= 0.336 N

b 1.6 m/s²

c the ball in part **b**; it has greater mass, so requires a bigger force to change its motion

d double it; if the mass doubles, you need to double the force to produce the same acceleration

P10.2

A force, matter

10

terminal

frictional, zero

B a mass is the quantity of matter, in kg,
      which is what bathroom scales show;
      weight is the force of gravity acting on
      a mass, in N, so the scales read mass not
      weight
   b 2 (N), 1.1 (kg), 0.053 (N)
C a 10 m/s²
   b there are forces other than gravity acting
      on the stone, e.g. water resistance
   c the velocity increases until it hits the
      water because the force of gravity is
      bigger than air resistance; when it hits
      the water, it slows down because of
      the force of the water on the stone; it
      reaches terminal velocity when the water
      resistance equals the weight
D straight diagonal line (or line with slightly
   decreasing gradient) until it hits the water;
   decrease in velocity to much smaller value
   with vertical line with decreasing gradient,
   then horizontal line, then vertical line when it
   hits the bottom; an 'x' on the horizontal line

P10.3

A driving
   thinking, braking
   braking, thinking
   N, mass (kg) × acceleration (m/s²)
B a braking distance: the distance a car
      travels while the driver is applying the
      brakes; depends on speed, mass of
      the car, brakes (force, condition), road
      condition
      thinking distance: the distance the car
      travels while the driver is reacting (i.e.
      during the reaction time); depends on
      speed, tiredness, level of alcohol in blood,
      distractions
   b e.g. if the lorry has a much slower speed
      than the car, but they have similar
      brakes / reaction times
C a force = 1000 kg × 4 m/s²
         = 4000 N = 4 kN
   b u = 30 m/s
      acceleration = $\frac{-u^2}{2s}$
         = $\frac{-900 \, m^2/s^2}{2 \times 120 \, m}$
         = −3.75 m/s²
                 (a deceleration)
      yes, because the deceleration of
      magnitude 4 m/s² is greater than this
   c if the acceleration is constant then the
      distance depends on the speed squared

P10.4

A kg m/s, mass, kg, × velocity, m/s
   the same as, conservation
B momentum = mass × velocity.
   if the mass of the lorry is twice that of the car
   it needs to have half the velocity to have the
   same momentum as the car
C momentum = mass × velocity
            = 300 000 kg × 2 m/s
            = 600 000 kg m/s

D a momentum before = momentum after
      0.3 kg × 2 m/s = 0.15 kg × velocity
      0.6 kg m/s = 0.15 kg × velocity
      velocity = 4 m/s to the right.
   b momentum is conserved / it is a closed
      system
   c either trolley X is moving faster, or trolley
      Y is moving slower; the momentum
      after the collision is not zero, but in the
      direction of trolley X

P10.5

A elastic
   difference
   directly proportional, proportionality, linear
   proportionality, non-linear, proportional
B a apply a force to the material and release
      it; an elastic material goes back to its
      original shape when you remove the
      force, but a plastic material does not
   b force and extension
C a extension = 4.5 cm − 3 cm
               = 1.5 cm = 0.015 m
      force = 40 N/m × 0.015 m
            = 0.6 N
   b if the force trebles, the extension trebles
      extension = 3 × 0.015 m = 0.045 m
      so length = 0.03 m + 0.045 m = 0.075 m
                = 7.5 cm
D a the first section of the graph is a straight
      line, so extension is proportional to force
   b sample would not return to its original
      length / it would be permanently
      stretched
   c two appropriate measurements used
      (e.g. 2 N / 2 cm, 4 N / 4 cm)
      spring constant = $\frac{force}{extension}$
                      = $\frac{4 \, N}{0.04 \, m}$
                      = 100 N/m

P10 Practice questions

01 speed affects the thinking distance [1]
    because you will travel further
    during your reaction time if your
    speed is higher; [1] speed affects the
    braking distance [1] because if you are
    travelling faster, the distance needed to
    stop whilst your brakes work will be longer;
    [1] stopping distance = thinking
    distance + braking distance [1] so your
    overall stopping distance will be larger [1]
02.1 0.0226 N, [1] 670 (m/s²), [1] 300 kg [1]
02.2 the mass in the column is the inertial mass
      because each mass = force/acceleration. [1]
03.1 weight / gravity [1]
03.2 force of the air on the ball/air resistance [1]
03.3 3 m/s [1]
03.4 it travels at a steady speed / terminal
      velocity [1] because the forces are
      balanced; [1] the air resistance is equal to
      the weight [1]

04.1 deceleration $a = \frac{-u^2}{2s}$ [1]
          = $\frac{-(60 \, m/s)^2}{(2 \times 50 \, m)}$ [1]
          = −36 m/s² [1]
04.2 F = ma [1]
      F = 250 000 kg × 36 m/s² [1]
        = 9 000 000 N [1]
04.3 spring constant = $\frac{force}{extension}$ [1]
          = $\frac{9\ 000\ 000 \, N}{10 \, m}$
          = 900 000 N/m [1]
04.4 momentum = mv [1]
          = 250 000 kg × 60 m/s
          = 15 000 000 kg m/s [1]
05   example answer: student A is increasing
      the force [1] so the acceleration increases;
      [1] the acceleration is not proportional to
      the force because the mass of the system
      is changing; [1] student B is also increasing
      the force, so the acceleration increases; [1]
      he keeps the mass of the system the same,
      [1] so the acceleration is proportional to
      the force [1]

P11.1

A information
   transverse, ripples, electromagnetic
   longitudinal, sound
   medium
B a move her hand at 90° to the direction of
      the rope
   b a longitudinal wave vibrates in the same
      direction as the wave travels; it is not
      possible to make a rope vibrate in
      this way
   c information or energy
C a wave B; the ground is moving at 90° to
      the direction of the wave
   b a compression is where the material
      is compressed / particles are closer
      together; a rarefaction is where the
      material is stretched / particles are
      further apart
   c they need a medium to travel through

P11.2

A maximum, peak, trough
   same, peak, peak, (or trough/trough)
   (s), $\frac{1}{frequency \, (Hz)}$
   (m/s), frequency (Hz) × wavelength (m)
B period = $\frac{1}{200 \, Hz}$
          = 0.005 s
C a speed = 660 Hz × 0.5 m
            = 330 m/s
   b 0.25 m

c $frequency = \dfrac{speed}{wavelength}$

$= \dfrac{4000 \text{ m/s}}{70 \text{ m}}$

$= 57 \text{ Hz}$

D a *The distance from one peak to the next peak is the wavelength. Frequency is the number of waves per second.* The unit of frequency is seconds. *The unit of wavelength is metres.* The distance from a peak to a trough is the amplitude.

b The unit of frequency is Hertz (Hz). The maximum displacement / distance of a peak or a trough from its rest / undisturbed position is the amplitude.

P11.3

A same, speed
refracted, zero
refraction
transmitted, absorbed

B a reflection: dashed lines drawn at equal angles, as in Figure 1 on page 140 of the Student Book; refraction: dashed lines parallel to the new direction, but further apart

b arrow top left continued until it hits the mirror, then a line drawn at equal angles; line at 90° to the mirror where the arrow hits the mirror; angle of incidence labelled from the normal to the incident ray; angle of reflection labelled from the normal to the angle of reflection

c the wavelength increases because the speed is increasing and the frequency is the same

d both waves change speed, both waves change wavelength; the wavelength of the waves in diagram **Y** increases because the speed increases; they are refracted because the direction changes; in diagram **Z**, the wavelength decreases because the speed decreases, but the direction does not change so they are not refracted

C light is transmitted through water from the bottom of a swimming pool, so you can see it; light is absorbed by very deep water so does not reach the bottom of a deep ocean

P11.4

A reflects
time, distance
m/s, $\dfrac{distance \text{ (m)}}{time \text{ (s)}}$

B a a reflection of sound

b measure a long distance from a large building using a measuring tape or a metre rule; stand that distance away; make a short loud sound using the blocks and start a stopwatch; stop the stopwatch when you hear the echo of the sound; multiply the distance by 2 to

use in the equation; use the equation $speed = \dfrac{distance}{time}$ to calculate the speed

c total distance travelled by the sound
$= 2 \times \text{distance to wall} = 2 \times 150 \text{ m}$
$= 300 \text{ m}$

$speed = \dfrac{300 \text{ m}}{0.90 \text{ s}}$

$= 333.3 \text{ m/s} = 330 \text{ m/s to 2 s.f.}$

d reasonable estimates, e.g. 5 cm, 0.2 s

e the biggest difference in the speed would be if the distance was 145 cm, or 1.45 m, and the time was 0.7 s, giving a speed of $145 \text{ m} \times \dfrac{2}{0.7 \text{ s}} = 414 \text{ m/s}$

P11 Practice questions

01.1 D [1] or F [1]; A [1] or C [1]

01.2 the direction of motion is at 90° to the direction of the wave [1]

02.1 mechanical waves need a medium to travel through, but electromagnetic waves do not [1]

02.2 electromagnetic waves have a very high speed / 300 million m/s, but mechanical waves travel much more slowly [1]

03.1 speed = frequency × wavelength [1]
$= 1000 \text{ Hz} \times 1.5 \text{ m}$ [1]
$= 1500 \text{ m/s}$ [1]

03.2 $period = \dfrac{1}{frequency}$ [1]

$= \dfrac{1}{1000} \text{ Hz}$ [1]

$= 0.001 \text{ s}$ [1]

04.1 diagram showing incoming wave crests and outgoing wave crests at 90° (correct by eye); [1] a barrier in an appropriate position (at 45° to the direction of incoming and outgoing wave direction) [1]

04.2 reflection [1]

04.3 transmission [1] and absorption [1]

05.1 they started the timer when they saw the sound being made [1] and stopped the timer when they heard it [1]

05.2 total distance travelled by the sound
$= 2 \times \text{distance to wall} = 2 \times 200 \text{ m}$
$= 400 \text{ m}$ [1]

$speed = \dfrac{400 \text{ m}}{1.3 \text{ s}}$ [1]

$= 307.7 \text{ m/s}$ [1] $= 310 \text{ m/s}$ [1]

05.3 repeat the experiment / take more measurements of time and calculate the mean time [1]

06.1 the depth of water [1]

06.2 the speed of the waves [1]

06.3 depth on the x-axis and speed on the y-axis; [1] a line that shows that as speed increases, depth decreases (straight or curved); [1] explanation: as the depth increases, the speed of the waves decreases [1]

P12.1

A microwave, infrared, visible light, ultraviolet, X-rays
longest, shortest, lowest, highest
visible light, 350, 650
source, detector
wave speed (m/s) = frequency (Hz) × wavelength (m)

B a radio waves
b radio / microwaves / infrared / visible
c gamma
d radio / microwaves / infrared
e any apart from infrared
f radio waves

C wavelength = 0.03 m
speed = 10 000 000 000 Hz × 0.03 m
$= 300 000 000 \text{ m/s}$

D biggest wavelength = 650 nm
$= 650 \times 10^{-9} \text{ m}$
so smallest frequency
$f = \dfrac{v}{\lambda}$

$= \dfrac{300 000 000 \text{ m/s}}{650 \times 10^{-9} \text{ m}} = 4.6 \times 10^{14} \text{ Hz}$

smallest wavelength = 350 nm
$= 350 \times 10^{-9} \text{ m}$
so biggest frequency
$f = \dfrac{v}{\lambda}$

$= \dfrac{300 000 000 \text{ m/s}}{350 \times 10^{-9} \text{ m}} = 8.6 \times 10^{14} \text{ Hz}$

P12.2

A waves
information, optical
radio, microwaves, microwaves, radio waves
radio, microwaves, infrared

B infrared cameras detect infrared radiation given out by objects, but remote controls emit it
remote controls use infrared radiation to transmit information, but infrared cameras use the infrared radiation they detect to produce an image, and so provide information

C a skin detects the heat from infrared waves, and eyes detect light waves
b point a remote control at a wall and see if you can still change channel
c you can detect microwaves using your mobile phone, and radio waves using a television or radio whilst inside your house

D a $\dfrac{2.45 \text{ GHz}}{0.1 \text{ GHz}} = 24.5$ (no unit)

b the energy of microwaves used in the microwave oven is much higher than the energy of microwaves used in the mobile phone (24.5 times higher); the frequency of microwaves for mobile phones needs to be so much lower, otherwise it could pose a risk to human health; the frequency of microwaves for microwave ovens needs to be high, otherwise it would not cook food

P12.3

A distance, spread, information, information,
amplitude
microwaves
evidence
visible, infrared

B a **X, Z, W, Y**
  b i unlike radio waves, microwaves
      do not spread out, and are not
      absorbed by the atmosphere.
    ii local radio stations use shorter
      wavelength radio waves because
      the range over which they need to
      be transmitted is not as far as those
      needed for national radio stations

C a infrared radiation is used to send signals
    down optical fibres
  b microwaves have a heating effect, but
    the infrared radiation is contained inside
    the optical fibres

D a **Y, X, Z**
  b a radio transmitter uses the audio
    signal to modulate the carrier wave,
    then transmits the modulated wave; a
    radio receiver receives the radio signal,
    separates the audio signal from the
    carrier wave, and emits the audio signal

P12.4

A shorter, skin
hospitals / medicine
bacteria, sterilise, cancer
ionising
damage

B a ultraviolet is less penetrating than X-rays
    or gamma waves, so does not penetrate
    further than the skin; it cannot damage
    cells inside the body to cause cancer
  b an X-ray or a gamma ray passes through
    human tissue; as it passes through, the
    X-ray or gamma ray knocks an electron
    out of an atom; this process is called
    ionisation; ionisation can kill cells or
    damage the DNA of a cell; damaged DNA
    can cause cancer
  c the shorter the wavelength, the higher
    the risk of damage to DNA
  d kill harmful bacteria on food / sterilise
    surgical equipment / kill cancer cells

C the Earth is protected by the ozone layer,
which absorbs ultraviolet radiation, just as
sunscreen does

D X-rays and gamma rays are different because
they have different wavelengths /
frequencies; X-rays are produced by fast-
moving electrons, but gamma rays are
produced in nuclear decay; X-rays and
gamma rays are similar because they are both
forms of ionising radiation, and they can both
cause cancer inside the body

P12.5

A cancer
ionising, low, high
absorb

B a produce images of internal body parts
  b kill skin cancer cells
  c kill cancer cells inside the body

C a to stop exposure to visible light
  b lines that end on the bone, but
    go through the gap in the middle;
    explanation: the bone absorbs the X-rays,
    so they do not reach the film, and it is
    not exposed; the skin / muscle does not
    absorb X-rays (as much) so they do reach
    the film; the image on the film shows
    areas that X-rays have reached
  c lead: it is very dense and absorbs X-rays

D a $\dfrac{2\text{ mSv}}{0.1\text{ mSv}} = 20$
  b a doctor might be more likely to ask for
    a chest X-ray because it gives a patient
    5% of their annual dose, but less likely to
    order a whole-body scan because it give
    a patient 5 times their annual dose; if a
    person's dose is higher, the risk of cancer
    is higher, so the condition being treated
    would need to be serious so that the
    benefits of the scan outweigh the risk

P12 Practice questions

01.1 radio, microwave, infrared, visible,
  ultraviolet, X-rays, gamma
  [radio before microwave, microwave before
  infrared, infrared before visible, visible
  before ultraviolet, ultraviolet before X-rays,
  X-rays before gamma – 1 mark each]

01.2 they are in order of increasing
  frequency [1]

01.3 visible light [1]

02.1 for example:
  1. microwave oven, [1] microwaves [1]
  2. remote control, [1] infrared [1]

02.2 for example:
  1. TV aerial, [1] radio waves [1]
  2. radio, [1] radio waves [1]

03.1 a high frequency electromagnetic wave
  used to carry information [1]

03.2 they carry more information / they do not
  spread out as much [1]

03.3 microwaves are used in a microwave
  oven to heat food, but microwaves are
  used in mobile phones to send / receive
  information [1]
  the mobile phone radiation does not
  produce a heating effect that can cook
  food [1]
  so it must have a smaller frequency /
  energy, and a bigger wavelength [1]

03.4 doing experiments on people involves
  ethical issues; [1] it is difficult to measure
  the heating effect / damage of microwave
  radiation on the brain [1]

04.1 radiation that removes electrons from
  atoms / molecules when it passes through
  matter [1]

04.2 in the parts that are white, the X-rays are
  absorbed by bone [1] so do not reach
  the CCD / photographic film, so no
  radiation = white; [1] the X-rays reach the
  CCD / photographic film through skin /
  muscle so radiation = black [1]

04.3 **two** from: shielding; [1] reducing time
  spent near source; [1] increasing distance
  between source and person [1]
  explanation: reduces damage to DNA from
  ionisation so reduces the risk of cancer [1]

05 wavelength = $\dfrac{\text{speed}}{\text{frequency}}$ [1]

  $= \dfrac{300\,000\,000 \text{ m/s}}{100\,000\,000 \text{ Hz}}$ [1]

  $= 3\,\text{m}$ [1]

06.1 X-rays are not absorbed (noticeably) by the
  kidneys [1]

06.2 when you have an X-ray, the source is
  outside the body, not inside; [1] so during
  an X-ray, the patient can be shielded with
  lead, but when a gamma camera is used,
  they cannot [1]

P13.1

A repel, attract
north, south
induced
steel
iron

B put the permanent magnet near one end of
the bar; turn the bar through 180°; if the bar is
only attracted to the magnet, then it is made
of a magnetic material; if one end of the bar
repels the magnet then it is a magnet itself

C a new paperclips are not magnetic / are
    not induced magnets
  b steel; when you put a steel paper clip in
    a magnetic field it becomes an induced
    magnet, and will attract other magnetic
    materials; steel is a hard magnetic
    material, and will stay magnetic when
    the magnetic field that made it magnetic
    is removed

D a

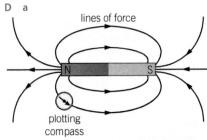

lines of force

plotting
compass

  b when the compass is near the magnet
    it will line up with the magnetic field of
    the magnet; as you move the compass
    away it will eventually line up with the
    magnetic field of the Earth

P13.2

A circles
parallel
increasing, reverse
parallel

B **a** concentric circles getting further apart; arrows to show field is clockwise

b the arrows would go in a anticlockwise direction

C **a** a solenoid is a coil of wire, and an electromagnet is produced when a current flows in the coil of wire

b the magnetic field inside an electromagnet is strong and uniform; the field lines are parallel inside; the field outside is looped like the field of a bar magnet, with the complete loops passing inside as well

D **a** the number of coils, the type of core

b a graph with a positive gradient (straight line)
y-axis labelled 'mass of iron filings'; x-axis labelled 'current'

c as you increase the current the strength of the electromagnet increases, so the mass of iron filings increases.

d advantage: the strength can be measured more accurately by finding mass
disadvantage: the paperclips are easier to remove from the electromagnet in order to count them

P13.3

A increase, reverse
coil
flux density
N, magnetic field strength (T) × current (A) × length (m)

B **a** the wire needs to be at right angles to the magnetic field

b $F = BIl$
$= 0.4\,T \times 2.0\,A \times 0.2\,m$
$= 0.16\,N$

c $I = \dfrac{F}{B \times l}$
$= \dfrac{0.05\,N}{0.4\,T \times 0.2\,m}$
$= 0.625\,A\ (0.63\,A)$

C **a** arrow starting on dot and pointing vertically up

b a coil of current-carrying wire in a magnetic field that can spin

c spin in the opposite direction; spin more slowly; spin faster

P13 Practice questions

01 like poles repel, [1] unlike poles attract [1]

02.1 the magnet does not need to be turned on and off; [1] the electromagnet would need a power supply and the wires would be inconvenient [1]

02.2 you would not be able to turn the electromagnet off [1] because steel is a hard magnetic material and keeps its magnetism when the magnetic field is removed [1]

03.1 the wire gets hot; [1] only connect the battery for short periods of time [1]

03.2 the current in wire **A** is into the plane of the paper, the current in wire **B** is out of the plane of the paper; [1] there is a bigger current in wire **B** than in wire **A** [1]

04.1 **B, A, C, D** [B before A – 1, A before C – 1, C before D – 1]

04.2 solenoid **B** has an iron core, but solenoid **A** has an air core; [1] when the iron core is in the magnetic field of the wire it becomes an induced magnet, [1] which means the magnetic field of the solenoid is equal to that of the coil + that of the iron, so the solenoid is stronger [1]

05.1 arrow on the long left side of the coil pointing up coil (towards commutator); [1] arrow on the long right side of the coil pointing down coil (away from commutator) [1]

05.2 the coil would spin the opposite way [1]

05.3 magnetic flux density
$= \dfrac{force}{length \times current}$ [1 – or implied]
$= \dfrac{0.005\,N}{0.05\,m \times 1.2\,A}$
$= 0.08$ [1] T [1]

Appendix 1: Physics equations

You should be able to remember and apply the following equations, using SI units, for your assessments.

Word equation	Symbol equation
weight = mass × gravitational field strength	$W = mg$
force applied to a spring = spring constant × extension	$F = ke$
acceleration = $\dfrac{\text{change in velocity}}{\text{time taken}}$	$a = \dfrac{\Delta v}{t}$
ⓗ momentum = mass × velocity	$p = mv$
gravitational potential energy = mass × gravitational field strength × height	$E_p = mgh$
power = $\dfrac{\text{work done}}{\text{time}}$	$P = \dfrac{W}{t}$
efficiency = $\dfrac{\text{useful power output}}{\text{total power output}}$	
charge flow = current × time	$Q = It$
power = potential difference × current	$P = VI$
energy transferred = power × time	$E = Pt$
density = $\dfrac{\text{mass}}{\text{volume}}$	$\rho = \dfrac{m}{V}$
work done = force × distance (along the line of action of the force)	$W = Fs$
distance travelled = speed × time	$s = vt$
resultant force = mass × acceleration	$F = ma$
kinetic energy = 0.5 × mass × (speed)²	$E_k = \dfrac{1}{2}mv^2$
power = $\dfrac{\text{energy transferred}}{\text{time}}$	$P = \dfrac{E}{t}$
efficiency = $\dfrac{\text{useful output energy transfer}}{\text{total input energy transfer}}$	
wave speed = frequency × wavelength	$v = f\lambda$
potential difference = current × resistance	$V = IR$
power = current² × resistance	$P = I^2R$
energy transferred = charge flow × potential difference	$E = QV$

You should be able to select and apply the following equations from the Physics equation sheet.

Word equation	Symbol equation
(final velocity)^2 – (initial velocity)^2 = 2 × acceleration × distance	$v^2 - u^2 = 2as$
elastic potential energy = 0.5 × spring constant × extension^2	$E_e = \frac{1}{2}ke^2$
period $= \dfrac{1}{\text{frequency}}$	
Ⓗ force on a conductor (at right angles to a magnetic field) carrying a current = magnetic flux density × current × length	$F = BIl$
change in thermal energy = mass × specific heat capacity × temperature change	$\Delta E = mc\Delta\theta$
thermal energy for a change of state = mass × specific latent heat	$E = mL$
Ⓗ potential difference across primary coil × current in primary coil = potential difference across secondary coil × current in secondary coil	$V_s I_s = V_p I_p$